Unsafe at Any Height

JOHN GODSON

 SIMON AND SCHUSTER • NEW YORK • 1970

All rights reserved
including the right of reproduction
in whole or in part in any form
Copyright © 1970 by John Godson
Published in the United States by Simon and Schuster
Rockefeller Center, 630 Fifth Avenue
New York, New York 10020

FIRST U.S. PRINTING

SBN 671-20708-3
Library of Congress Catalog Card Number: 70-130475
Designed by Eve Metz
Manufactured in the United States of America
By American Book–Stratford Press, Inc., New York

To ARTHUR, THELMA, PETER, ISOBEL, TONY and CHRISTOPHER LITTLE, without whose kindness and friendship this book might never have existed. It was they who awoke in me the spirit of adventure and the yearning for new experiences through world travel.

CONTENTS

	Foreword	9
One	"Sir, do you think we're going to get out of here?"	15
Two	The Man in the Cockpit	40
Three	The Overcrowded Airports	68
Four	The Passenger's View	89
Five	Profits versus Safety	114
Six	The Charter Game	151
Seven	Designed for Disaster	164
Eight	What to Do in an Emergency	189

Foreword

Aviation accidents are something which most of us take for granted. Newspaper reports of a disaster arouse sympathy, but people seldom ask themselves whether it need have happened at all. Several months later a summary of the findings of the board of inquiry will be published and may reawaken brief interest. But boards of inquiry rarely look deeply enough into the secondary causes of an accident, and it is even more rare for boards to make recommendations for the avoidance of such occurrences in the future. In general, the public believes that the annual toll of deaths and injuries in air crashes is unavoidable.

This book attempts to show that this belief is a myth. It has been written to demonstrate also that standards of air safety have been, for years, culpably neglected. I look to the past mainly to emphasize the lessons for the present and the future; to show that something can and must be done to improve air safety before we move unthinkingly into the age of the giant passenger aircraft and the supersonics.

My own interest in the subject began several years ago. I was struck by the fact that major accidents could happen for the most trivial and haphazard reasons. It was puzzling, and as I am an inveterate air traveler, my curiosity was aroused. This was the beginning of several years of research into the subject, a pursuit not without its difficulties. I found that no single country's aviation authorities appeared to have complete sets of accident records. But eventually—largely with the help of the International Civil Aviation Organization—I collected most of the official reports of all accidents which happened in world aviation since 1949, a total of some fifteen thousand mishaps.

Reading this massive file, I found a disturbing pattern emerging. The same kinds of accidents repeated themselves over and over again, and for the same reasons. The board of inquiry in each case had brought in its tentative verdict, and that, all too often, seemed to be the end of the affair—until the next accident and the next inquiry. The absence of any sign of progress, or even any great interest in the subject on the part of the aviation industry, was one of the most conspicuous failings. A further pattern emerged when I was making a compilation of the accidents and incidents relative to each of the air companies. I could see that some airlines were frequently involved in misadventures, while other companies had persistently better records. Why can some airlines amply demonstrate their interest in passenger safety while others are regularly cited in the accident lists?

Talks with a large number of people connected with aviation and air safety only confirmed my view that safety standards were several years out of date, and that lives were being lost each year because of this. This book is written not merely to inform the reader, but in the hope that some action will result from its publication.

I would like first to thank the dozens of people who for obvious reasons cannot be named—pilots, flight engineers, design officers, airline cabin and maintenance staff—the people who are fully aware of what is going on but whose jobs would be in jeopardy if their names were to appear in this book.

I would also like to thank the following persons and organizations for the tremendous assistance they gave me in researching this subject, *with the proviso that the views and feelings expressed in this book do not necessarily rep-*

resent those of the people and companies listed:

Bob Bennett, Laurie Taylor and Ralph Kohn, of the British Airline Pilots Association; Beverly Rowe, of the London Computer Centre; Alan E. Smith, of British Aviation Insurance Ltd.; Kenneth L. Burroughs, of the (American) Air Line Pilots Association; Edward J. Slattery, of the National Transportation Safety Board; A. M. Lester, Joachim Schypek and Sydney Cooper, of the International Civil Aviation Organization; Donald Pengelly and Dr. Richard Shaw, of the International Air Transport Association; Charles G. Warnick and Eugene Kropf, of the Federal Aviation Agency; Benjamin Cook, William Miller and Thomas Laughlin, of the Lockheed Aircraft Corporation; Crosby Maynard, Hu Gagos and Charles Strang, of the McDonnell Douglas Corporation; William Worden, Bill Purdy and Gordon Williams, of the Boeing Company; and B. W. Townshend, aircraft ditching specialist.

Thanks to the following organizations for assistance with statistical information:

International Civil Aviation Organization; Air Line Pilots Association; Civil Aeronautics Board; Lloyd's Insurance of London; American Insurance Association; the chargés d'affaires in London of Canada, Ireland, Spain, Portugal, France, West Germany, Switzerland, Austria, Italy, Argentina, Japan, and Australia; Canadian Department of Civil Aviation; Australian Department of Civil Aviation; Japanese Department of Civil Aviation; *Flight International* Magazine; United States Army Aerial Research Unit; Federal Aviation Agency; National Transportation Safety Board.

Appreciation for assistance in technical matters goes to the (American) Air Line Pilots Association; British Airline Pilots Association; Australian Department of Civil Aviation; United States Department of Defense; United States Department of Commerce; National Transportation Safety Board; Federal Aviation Agency; Civil Aeronautics Board of the United States; National Aeronautics and Space Administration; Flight Safety Foundation; United States Army Aerial Research Unit; Air Safety Committee of the United Kingdom; The Boeing Company; McDonnell Douglas Corporation; Lockheed California Corporation; Barnes Engineering Company, Inc.; Hardman Engineering Company; Canadian Civil Aviation Department; International Civil Aviation Organization; Explosive Technology, Inc.; Flight Standards Service; Shell Oil Company; Esso Petroleum Ltd.; United States Army Aviation Material Laboratories; Airworthiness and Performance Committee; Aeromedical Coordinating Committee; Destructive Objects Detection Committee; Rolls-Royce Ltd.; Pratt & Whitney Corporation; Hawker Siddeley Ltd.; Dunlop Rubber Company; International Air Transport Association; Sud Aviacion; S.N.P.L. France; International Federation of Air Line Pilots Associations; *Flight International* Magazine; the *Aeroplane Magazine;* Lloyd's of London Aircraft Accident Lists.

Very limited assistance with much delay and argument was gained from Air Registration Board of the United Kingdom; Board of Trade, Civil Aviation Department of the United Kingdom; British Airports Authority.

Finally, a word about the word *probable*. You are going to have to wade through a great number of *probables* in this book. This is because up to now, and even now, boards of aircraft accident investigation have recorded only the "probable causes" of a disaster, mishap or accident. The reasons for this policy are long and complicated and are discussed in the text later on.

To put it briefly, they do this probably because in certain cases it may be difficult to discover the real causes; probably because they may want to cover up some major governmental blunder that could have been a contributing cause; or possibly to save litigation. It is with this last-mentioned thought in mind that we too must include the "probables."

List of Abbreviations

A.L.P.A.	Air Line Pilots Association (American)
A.R.B.	Air Registration Board (of the United Kingdom)
A.T.C.	Air Traffic Control
B.A.L.P.A.	British Air Line Pilots Association
B.E.A.	British European Airways Corporation
B.O.A.C.	British Overseas Airways Corporation
C.A.B.	Civil Aeronautics Board
CAP	Civil Air Publication (British)
CAT	"clear air turbulence"
F.A.A.	Federal Aviation Agency
I.A.T.A.	International Air Transport Association
I.C.A.O.	International Civil Aviation Organization
IFALPA	International Federation of Air Line Pilots Associations
ILS	"instrument-landing system"
JAL	Japan Air Lines
JP4	"Jet Power #4"; a kind of fuel—"wide-cut gasoline"
N.T.S.B.	National Transportation Safety Board
SNECMA	French Aviation Equipment Manufacturer
S.N.P.L.	Société Nationale Pilote de Ligne (France)
TWA	Trans World Airlines
USAAVLABS	U.S. Army Aviation Laboratories
VASIS	Visual Approach Slope Indicator System

1 · "Sir, do you think we're going to get out of here?"

On a clear but chilly November evening in 1965 a Boeing 727 took off from Denver airport on its regular scheduled flight to Salt Lake City, Utah.

Aboard the aircraft everything proceeded in a quiet and normal way. In the left pilot's seat was Captain Gale C. Kehmeier, who had been with United Air Lines for twenty-four years and had accumulated over seventeen thousand hours of pilot time, including three hundred and thirty-four in the B 727, a relatively new aircraft.

On his right sat First Officer Philip E. Spicer, who was flying the aircraft under the captain's direction. He had over six thousand hours to his credit, of which eighty-four had been in the 727. Behind them, at the small navigation desk, was Second Officer Ronald R. Christensen. The weather reports presented no special problem; there was every reason to expect an untroubled flight. At 31,000 feet the aircraft leveled out at its cruising altitude.

Though the flight had originated in New York that morning, most of the eighty-five passengers had come aboard at Denver or one of the several intermediate stops.

As a group, they were more or less typical of one you might meet on a short-haul flight across the American continent. There were salesmen going west on a trip, businessmen heading back to San Francisco after a visit to branch offices, a couple of young Air Force men going home on leave, and a few elderly couples who were on their way to visit friends or relatives.

Many were experienced air travelers, like the three inspectors of the Federal Aviation Agency, sitting down the aisle, on their way to deal with some Agency business. Nearly all were strangers to one another, with a chance to make only the brief conversational links allowed by fifty-seven minutes' flying time.

The three stewardesses, Victoria Cole, Faye Johns and Annette Folz, went through their routine duties in the cabin. Some of the passengers passed the time by looking at the safety pamphlet in the seat pocket. Some, especially the inexperienced ones, gave the pamphlet some attention. One passenger, unusually conscientious, sitting by a right-wing emergency exit, made a mental note of the fact that the seat back in front of him might obstruct quick removal of the exit panel. He worked out a mental drill for doing it: Pull down handle, lift panel inward and tilt on edge, then push it out.

Some passengers had struck up sociable small talk with their neighbors. One was giving advice on hotels in Salt Lake; another was talking about his business in San Francisco; two others were discussing the merits of the 727. Asked later to try to remember who had been sitting in the seats near their own, most passengers could only recall their neighbors as anonymous details— ". . . man dandling child on knee"; ". . . heavy-set man in dark suit"; ". . . middle-aged lady"; et cetera.

When the flight was about seventy miles east of Salt Lake City, the captain was given permission by air-route traffic control to begin his descent down to 16,000 feet. It was arranged that the flight would come in from the south to land on Runway 34 Left, having fixed its position by the Lehi beacon southeast of the city. "Okay, we'll start her down," the controller heard. Ground radar followed the aircraft all the way along its descending path. Below 16,000 feet the flight entered a thick layer of overcast cloud which

17 "Sir, do you think we're going to get out of here?"

shrouded visibility and gave the aircraft a slight jolt or two from air turbulence.

At 1748 hours with about ten minutes to go before expected touchdown, the flight reported to the Salt Lake City Approach Controller, "Okay, we're slowed to two fifty [knots] and we're at ten [10,000 feet]; we have the runway in sight now, we'll cancel [that is, cancel instrument flying] and stand by with you for traffic." Passengers peering through the cabin windows could now see the lights of the city ahead; those who knew the place well could identify the bright lines of the main streets.

The landing gear was lowered and wing flaps were extended as the aircraft continued to reduce speed. Some of the passengers got the impression that they were coming in at an unusually steep angle. The descent was, in fact, being made at approximately two thousand feet a minute—an approach rate roughly three times steeper than that recommended to their pilots by United Air Lines.

At 1749 the aircraft was given permission to land. At about this moment (versions differ as to the precise timing) the first officer began to apply more power, possibly suspecting that the aircraft might land short of the runway. But the captain advised him to wait. About half a minute later the first officer moved the thrust levers halfway. There seemed to be no response from the engines. The captain pushed the levers to the full takeoff-power position and took charge of the controls. Seconds later, at 1752, the tower controller reported, "United's on fire . . . just landed . . ."

The aircraft's first impact with the ground occurred 335 feet short of the threshold of the runway.[1] The passengers felt it in various degrees according to their seat positions.[2] Some felt it as no more than the extra-heavy bump of a bad

[1] C.A.B. Accident Investigation Report. File No. 1–0032, June 1966.
[2] Accident Investigation Statements of Survivors, N.T.S.B., SA388 Exhibit No. 6.

landing. Others felt it much more violently. One man, sitting over the undercarriage could "feel and hear the [landing] gear ripping off the aircraft and the crunching of it scraping on its belly." The aircraft continued sliding on its fuselage for about half a mile, eventually swinging off to the right of the runway. One engine broke away as it swung and hurtled a hundred and forty feet on its own.

Despite the violence of the impact, apparently none of the crew or passengers was injured by it. All were still left with a chance to escape. But speed was vital. The difference between death and survival was compressed into that first minute or two. Everything depended on passengers' quickness of reaction and how well the emergency routines and the safety equipment worked.

A leg of the broken landing gear had pushed a large hole into the rear right-hand side of the fuselage, rupturing fuel lines situated a few inches inside. Fire broke out almost instantly, caused either by a spark from the scraping of metal on the runway or from severed generator leads. Flames burst through the fuselage where the landing-gear leg had penetrated and were licking menacingly at the floor below the passenger seats. The cabin began to fill with choking clouds of smoke. The lights flickered briefly then went out, adding darkness to the confusion.

In those first bewildering moments as the aircraft skidded along, one of the stewardesses shouted to the passengers to keep their seats until it stopped. Most did so, possibly in obedience to her instructions. But it is more likely that they were simply too shocked and helpless to move, unsure of which way survival lay.

Some of those in the forward part of the cabin seemed unaware of the blaze aft, until someone shouted, "My God, we're on fire. Let's get out of here." Some passengers were already scrambling to find the handle of the nearest emergency exit, or groping forward along the aisle toward the

19 "Sir, do you think we're going to get out of here?"

front door. A stewardess was struggling with the handle, hampered by the press of people behind her, and it was not until the second officer came back to help that they got it open.

By then the queue had been waiting for ten or fifteen seconds. "It seemed like an age," said one survivor later. Another, who escaped less than a minute after the crash, estimated that he would have succumbed to the smoke and heat in a few more seconds. Those out first remembered little jostling and few voices raised, possibly because everyone was trying to save every gasp of breath he could. Those who died were suffocated. This was shown by the high carboxyhemoglobin concentration in the victims.[3]

Preliminary indications showed that the interior furnishings accelerated the spread of the fire, contributing to the heavy black smoke, and hence to the fatalities. One man, who did not get out until the flames were running the length of the cabin ceiling, recalled that "it was impossible to get near any of the window exits, due to the crowding of the people . . . practically everyone was hysterical, with a great deal of screaming and shouting."[4]

The man seated by the starboard-wing exit, who had taken the trouble to work out a hatch-opening routine, was one of the first out. He got the hatch open quickly, while the aircraft was still moving, but was blown flat on his back as soon as he got the panel out. He struggled up again, went head first through the hatch onto the wing and escaped with minor burns and bruises.

Once out, survivors went on reacting almost unthinkingly. Some ran away as fast as they could. Some hovered in a daze by the aircraft, although dimly aware of the con-

[3] Letter from Director Bureau of Safety to George S. Moore, of the F.A.A., Dec. 16, 1965.
[4] Accident Investigation Statements of Survivors, N.T.S.B., SA388 Exhibit No. 6.

siderable risk of explosion; a few of these were sprawled on the ground, unable to move because of broken or twisted ankles or near-asphyxiation. One passenger recalled with bewilderment how he had run away, taken off his jacket, folded it with extreme neatness and care, and placed it on a clean patch of grass. Only then had he returned to help. One of the F.A.A. inspectors bravely crawled back up the safety chute to try to rescue a passenger. Seeing it was sheer suicide others pulled him back.

Within a few minutes at most, watchers outside thought it impossible that anyone should still be living in the inferno. The airport crash trucks arrived within three and a half minutes, but, since the fire was largely internal, the hoses were not fully effective. One survivor claimed that the first truck had run out of foam before the others arrived.

At least three people remained alive aboard the aircraft. It is possibly the most astonishing example of survival in the air-accident records. One of this fortunate trio was a stewardess, Annette Folz. In her testimony to the Civil Aeronautics Board inquiry,[5] she described how, after the impact, she managed to open the rear (aft) pressure door. This leads to another "access door" and a ventral stairway, which under normal circumstances is lowered hydraulically to allow passengers to board and alight at the rear of the plane. Initially Miss Folz could not open the access door, because her hands were too badly burned. (Her hair had caught fire and she had had to put out the fire with her hands.) She found two men crouched on the stairs. The stairway was jammed almost against the ground outside, and there appeared to be no escape that way. So Miss Folz turned back to the cabin, intending to try to help evacuation at another exit. By then most of the cabin was on fire

[5] My narrative of the crash follows the facts in the official inquiry report (C.A.B. File No. 1–0032 and SA388 Exhibit No. 6).

21 "Sir, do you think we're going to get out of here?"

and she could see that it would have been "sure suicide" to try that way. "I remembered what we were told in training school, 'If there's a solid block of fire in front of you, don't go through it.'"

The two men were at the very end of the stairs laying down. I curled up right behind them into a little ball to get away from most of the smoke and fire and I started breathing through my jacket. I couldn't see any possible means of escape for us. I thought for sure that I was going to die. I then began to pray and review my life. It was getting hotter and hotter on the stairwell. The two men weren't talking so I asked one, "Sir, do you think we're going to get out of here?" His answer was "Yes." I couldn't believe it, so I asked, "Alive?" He then said "Yes" again. It was wonderful to see that he was optimistic.

By then the fire was getting closer and the aft pressure door itself was alight. Miss Folz realized that if a spark fell on her nylons "my whole legs would go up in flames." So, after a painful struggle, she took them off. She encouraged the two men to make as much noise as possible to attract the firemen's attention. Then they noticed a two-inch crack in the fuselage through which their vital supply of fresh air was coming. Miss Folz put her hand through it and waved. The firemen saw it, directed foam on the tail, and passed another hose inside for one of the men to use. The firemen got them out at last, after they had survived between twenty-five and thirty minutes on the burning plane.

All together, forty-three people, all of them passengers, died in the Salt Lake City crash.

The main conclusions of the Civil Aeronautics Board inquiry were that the aircraft had crossed the outer airport marker more than two thousand feet above the proper glide slope; that power had been applied too late to arrest an excessive rate of descent; and that the captain's train-

22 Unsafe at Any Height

ing records indicated a tendency to deviate from acceptable standards and tolerances. Captain Kehmeier disagreed with these findings and with the implication of "pilot error." Although the C.A.B. gave the B 727 a clean bill of health, it was puzzling that, within a period of nine months, four aircraft of this type had crashed in the course of a landing approach, all with heavy death tolls and all in good weather.[6]

Though "obvious," the lessons which could have been learned from countless accidents and boards of inquiry have not yet been put imaginatively into practice in civil aviation. The Salt Lake City accident is a classic case of a "survivable" accident—that is, one in which all of the people aboard lived through the initial impact and might have been saved if the means had been available. And since it is a relatively recent accident, which involved an aircraft regarded as one of the high-performance models of the late 1960s, it will be clear that the safety standards I discuss are contemporary. The Salt Lake crash is unusual because, although it was a disaster of some magnitude, it left enough survivors to give us a clear picture of what happened during those vital few seconds in which everyone's survival was in the balance.

A number of the survivors, in their official testimony, commented on various shortcomings in the safety equipment as they had *experienced* it, and suggested improvements which could have saved lives. These included the following:

1. More emergency exits with good, crashproof illumination.
2. A member of the cabin crew should be seated be-

---

[6] United Air Lines, Lake Michigan, Aug. 16, 1965, 30 dead; American Air Lines, Cincinnati, Nov. 8, 1965, 58 dead, 4 injured; United Air Lines, Salt Lake City, Nov. 11, 1965, 43 dead, 35 injured, 13 unhurt; All Nippon Airways, Tokyo, Feb. 4, 1966, 133 dead.

side each exit during takeoff and landing, ready to operate it.
3. Passengers should be given specific instructions before takeoff on how to open exits and manipulate door levers.

As one survivor pointed out, there is no assurance whatever under the present haphazard escape system that the passengers sitting next to an emergency exit (and thus theoretically entrusted with the job of getting it open after a crash) will not be elderly ladies, or foreigners who cannot read the instructions. Others suggested that a block of seats should be taken out to give better access to the exits, and that the inflation of escape chutes should be automatic. Others wondered why the fabric of the plane burned so quickly, and why the fumes were so toxic. They suggested that the fireproofing could be much improved.

It is important to remember that we are discussing survival problems in a contemporary aircraft whose safety standards, broadly speaking, are no better and no worse than those of any modern aircraft you may be flying in tomorrow. The points made by the survivors have often been made by safety organizations over the last decade and more.

Yet if that accident at Salt Lake were repeated tomorrow, either in a B 727 or almost any other jet aircraft now in operation, it is a near certainty that the passengers would be faced with the same survival problems.

After such a heavy landing there would be the same risk of fire. The aviation industry has not yet devised any adequate protection for the fuel lines or fuel storage tanks, should the undercarriage collapse. Because of the insufficient effort put into fire-prevention research over the past decade, the passenger would still have to struggle against the effects of toxic fumes, and flames would probably take hold just as quickly.

Neither emergency exits nor escape chutes have improved since Salt Lake. It would still be a rarity to find any special lighting indicating the position of these exits. The chances would be at least ten to one against finding a stewardess sitting beside each exit on landing and takeoff: a passenger's speed of escape would still depend on the fumbling efforts of the passenger nearest to his exit.

The airport fire brigade would no doubt show the same promptness in answering the emergency call as they did at Salt Lake, but with the methods and material available they would probably not get the fire out any quicker or be able to rescue any more passengers now than they did then.

It is true that in the United States the Federal Aviation Agency had updated its rules about evacuation speed.[7] A manufacturer must show (but only in a factory test) that evacuation can be achieved at roughly twice the rate previously required. This change is a step in the right direction. But experience shows that there is still a deplorably wide gap between the relatively progressive thinking of the F.A.A. and operational practice. And the situation tends to be a little worse outside the United States. In practice, the passenger would find that there had been virtually no improvement in the situation since Salt Lake City.

I have purposely confined myself here to noting a few of the defects in the passenger-cabin arrangements, since these can be seen and confirmed by any reader on his or her next flight. But air safety, of course, embraces a much wider range. In later chapters I shall discuss the need for cockpit instruments which would really help the pilot in today's flying conditions; for aircraft designed for safety as well as speed; and for airports which do not lay all the burden of being "right first time" on a tired flight crew.

Unhappily, one need only look at the accident records to

[7] Aeronautics and Space—Part 25—F.A.A. Federal Register 32. F.R. 13255, Sept. 20, 1964.

see how tragically slow the aviation industry has been to learn from experience. One need look through no more than ten years of records to see how certain accident situations repeat themselves.[8]

In fact, to describe them as accidents would, in the majority of cases, merely be a way of putting a comfortable gloss on the facts, in much the same way as the airlines do. There are, for instance, instruments so designed that they confuse pilots, seats that uproot themselves after only a moderate and otherwise survivable collision, airports which leave piles of bricks and other obstructions near the runways, airlines (including the most famous) which do not observe their own regulations about the stowing of carry-on luggage, charter companies which overload their aircraft or overwork their crews and persistently get away with it until a crash occurs, companies which carry out improper maintenance routines for long periods without detection by the government inspectorate and escape with only a mild reprimand when discovered.

Air accidents are broadly of two kinds. There are those which are more or less unpredictable, which result from fluke situations or from fatal defects that cannot be detected, and those wholly attributable to gross human and/or airline negligence. The early Comet crashes, caused by unsuspected metal fatigue, clearly fall into the first category.

It would appear that the crash of the British Midland Airways Argonaut aircraft in the middle of Stockport, England, in June 1967, with the loss of seventy-two lives, was an example of the second category.

Some incidental details of the Stockport crash are well worth recording, and I have done this by simply reproducing some of the statements of witnesses from the official accident report. First, the circumstances of the crash:

[8] I.C.A.O. Air Accident Digests.

At 1008 hours came the radio call reporting the plane in difficulty and the crash crew at the airport were alerted. Two minutes later the plane plunged into the centre of the town. It fell on Hopes Carr, one of the very few open spaces in the heavily built-up town centre, avoiding Stockport Infirmary, the new police headquarters and 3 multi-storey blocks of flats, which ringed the area of the crash. As the plane 'bellied' into the open patch it hit a 2 storey building, and landed on a garage which contained 12 vans. Impact damage was caused to an electricity sub-station.

The plane itself broke into 3 parts. The cockpit landed on the forecourt within 10 feet of two 500 gallon petrol tanks, each almost full, and two electric delivery pumps; the main fuselage with the passenger cabin hit the electricity substation overlooking Tin Brook, and the tail piece dropped on to the steeply sloping banks of the ravine. Shortly afterwards the plane burst into flames.

The sub-station contained two transformers with an input of 6,600 volts. These in turn caught fire and the electricity supply failed in a large area.

I reproduce the following extract for the light it throws on the difficulties of escape and rescue work. The questioner, a lawyer, is interrogating a policeman who was one of the first persons on the scene.

Q: When you were tackling the work of rescue were you able to undo the belts of the passengers?
A: I had some difficulty in undoing the belts of the passengers, as indeed one of the passengers was having difficulty in getting it off. In fact, as I went to the lady she was strapped in and she said to me "I can't get my seat belt off."
Q: You managed to get one lady free, and then there was a second lady, a middle-aged lady, that you tried to free, and again you had difficulty in undoing her safety belt?
A: I did, sir, yes.
Q: Because up until that time you had never had occasion to undo a safety belt before?
A: Not of this type, sir, no.

27 "Sir, do you think we're going to get out of here?"

Describing later events, the witness continued:

There was then a large explosion from the rear section of the aeroplane which prevented me from going to that area, where I could hear people screaming and shouting. The fire then flared up and I was driven back. I then assisted in extricating the co-pilot from the right-hand side in the cockpit. He was badly injured.

Thus, it appears that if the passengers could have been extricated more speedily from their seat belts, then this explosion would not have trapped so many who were still alive. According to Police Constable Oliver, another factor which complicated his rescue work and impeded passengers' escape was the seating, which was torn away from its mountings.

At this time it was dead quiet and all the passengers appeared to be seated, and facing towards the front of the plane, the seats appeared to have concertina-ed from the rear of the plane.

There did not appear to be any gangway between them and they appeared to be roughly in rows but all across the floor of the plane, I saw that some of them were obviously alive, and began to pull them clear of the plane. I do not know how many I removed but I remember undoing their seat belts (which I think were green).

The operation of removing people was made difficult by their legs being trapped under their own seats, which had broken away from their mountings, and on which the frames of the seats had become twisted. I remember them carrying a small boy, who I think was dead. Throughout this time I did not consciously remember going away from the plane, but passed survivors to some other person who took them away behind me, and about this time the other police arrived, amongst them I can remember . . .

The most important point that emerges is that it was the strengthening bars under the seats which in some cases trapped the passengers' legs.

Even more telling is the following extract from a statement made by another eyewitness to the crash.

Through the gaping hole that appeared in the main plane, immediately behind the port-side wing, one could see that passengers had been thrust forward into a heap. Some were conscious—others unconscious. Most were still strapped to seats that had apparently broken away from their moorings. Some appeared to be badly injured and some were partially stripped of clothing.

Clearly there could have been considerably less injury if the seats had been strengthened in some other fashion, preferably to the floor.

It was stated at the inquiry,[9] that engine failure (the main cause of the crash) could occur in this aircraft through fuel starvation if the fuel cocks were not fully switched to the "on" position; it also emerged that it was impossible for a short pilot to achieve this proper switch-on when strapped in his seat. Yet this type of aircraft had been in service for twenty years; this particular plane had passed numerous examinations of airworthiness and had been involved in at least two similar incidents, the reasons for which were officially "unknown."[10]

It is symptomatic of the history of air safety that in all these years not one inspector or government authority was prepared to insist that the Argonauts should be grounded until their absurd cockpit layout and the flight manuals (which omitted all mention of the peculiarities of the fuel system) were corrected. Moreover, the inquiry was told that there were nearly nine hundred aircraft in operation in late 1967 with fuel systems very similar to, or identical with, that of the Argonaut.

[9] British Board of Trade Report CAP 302.
[10] British Midlands Argonaut G-ALHG—May 14, 1967; British Midlands Argonaut G-ALHG—May 28, 1967.

29   "Sir, do you think we're going to get out of here?"

The board of inquiry pointed out that if there had been some reasonable exchange of information between companies running Argonauts and similar types, the defect would have been discovered much earlier. The fact that there had not been any such sharing of information about problems affecting passengers' lives as late as 1967 (and the situation is little changed) suggests to me the most culpable sort of commercial separatism. It seems to make nonsense of such bodies as the International Civil Aviation Organization (I.C.A.O.). It is known that this body meets to agree on conditions and regulations of international civil-aviation operations. If it cannot also work out a cooperative system which places some importance on safety, then the aviation industry is in a sad state. Nor is the I.A.T.A. (International Air Transport Association) much more concerned about safety. Instead, it seems bent on maintaining restrictive practices by setting fares for most of the world's airlines (a trick that would be against all antitrust and monopoly laws were the airlines dealing in chocolates instead of air travel).

The airlines and others responsible for aviation safety usually have a stock reply when their standards are challenged. It can be summarized thus: "Perfection can never be achieved. Air safety will always be a necessary compromise with expense and practicability."

A totally accident-free aviation world is, of course, not expected, and to use the word "perfection" is misleading. One is concerned about those many areas of civil air transport where neglect and inadequacy are glossed over and condoned by inaction year after year. The other misleading word is "compromise." As in any other commercial enterprise, a balance had to be struck between the service offered and the price the customer is prepared to pay. But it has to be remembered that here we are dealing not with merchandise but with human lives; and a company which takes part in this kind of business must expect to have to observe the most rigorous standards. Yet how does this "compromise" work in practice? It is striking how wide is the gap between the complacent aeronautical companies' safety-minded image of themselves, and the reality of their achievements. If there is a conflict between safety standards and expenditure—whether it is on the provision of life rafts, the best cockpit instrumentation, or safe passenger seating—it is disturbing to see how frequently economy wins. That is how "compromise" is interpreted.

Another way in which the industry answers criticism is by asserting that the accident rate has been steadily declining and that air travel is demonstrably safer than road transport, thus implying that the present number of air-passenger fatalities and injuries are, in a sense, "reasonable."

Overreliance on statistics is one of the vices of our age; the tables appear to show a gradual decline in fatalities to an average rate for the late 1960s of about 0.6 deaths for each million passenger-route-miles flown, but it is not clear what effect extraneous factors, which have little or nothing to do with the airlines' safety effort, have had on the figures. Because of the gradual familiarization of crews and operators with a variety of jet aircraft over the decade one would expect to see early problems being ironed out.

All one can say is that flying has increased year by year, and the number of fatalities and injuries also increases each year, but at a slower rate. As a current I.C.A.O. bulletin somewhat complacently put it, "The accident rate [for scheduled airline flights] for the late 1960s seems to have settled down to a figure of around seven hundred and fifty deaths per year."

Comparison with the road accident rate is also misleading, since the two sets of figures are literally noncomparable in any meaningful sense. The chances of road collisions, with so many millions more vehicles traveling at potentially lethal

"Sir, do you think we're going to get out of here?"

speeds in opposite directions, along narrow route ways, are vastly higher. We normally have the opportunity to choose whether we will drive or travel in a particular road vehicle whose defects, if any, will normally be self-evident. But once aboard an aircraft we are, in practice, utterly dependent on the conscientiousness with which an operator has prepared for safety.

Remembering that about a further seven hundred and fifty people are killed in "nonscheduled" and other flights, is seven hundred and fifty deaths a year on scheduled airlines truly an adequate standard? If we agree that aviation must set its own levels of care and concern and not conceal the issue by bogus comparisons with quite different forms of transport, then I believe that on the evidence it certainly is not.

From a close study of the accident investigations, covering some hundreds of crashes, ranging from major disasters to minor mishaps, it is possible to make a reasoned appraisal of what goes wrong and how far neglect is involved. On this evidence alone I would estimate that the air-accident rate could be reduced by half or possibly by more.

This could be done without demanding inhuman effort or extravagant expense and without setting unrealistic targets. In addition, the number of occasions when an aircraft comes near the brink of a crash situation, without quite going over the edge (incidents the average passenger seldom hears about, though they are much more common than supposed) could be drastically reduced.

The urgent need for a new spirit in tackling this problem cannot be too much emphasized. We are on the verge of an era in which there will be a fair number of giant jets in the sky, each carrying from three hundred to four hundred passengers. These leviathans will be operated by today's airlines and, unless a change of attitude can be quickly stimulated, with the same standards of safety which give such a hollow satisfaction to the statistics men today; a disaster involving one or more of these planes a year would shake the public confidence in aviation to the roots. Now is obviously the time to insist that the industry do something about it.

The flight usually goes so smoothly—so far as we are aware in the passenger cabin—that we scarcely give a thought to the problems of the pilot. Airline advertising has diverted our attention away from what happens on the flight deck, and it encourages us to think about the five-star dinner and the hostess service. This is all very understandable and, up to a point, quite proper. But complacency about the realities of air safety has now become so ingrained that it is time to redress the balance. Even the airlines now appear to have fallen victim to their own propaganda.

This situation continues largely because the aircraft-using public does not have access to the information that would stir it to protest. As an article in *Flight International* magazine (August 17, 1968) puts it: "Public opinion plays little part in air safety until a catastrophe happens, when there is a wave of (probably misinformed) indignation at the time, followed by a backwash when the inquiry report is published. Indignation is unleashed at the authorities, rather than at individual airlines . . ."

A crash is a crash: nothing can disguise it as anything but a tragedy. But the public will continue to get only the most superficial picture of the safety problem so long as they have to judge it in terms of the occasional big disasters that really make the headlines. What about the accidents that did not *quite* happen, the ones in which some things went wrong, but left the pilot just enough margin to save the day?

Such incidents are written off as part of the "normal" accident rate which the average airline is willing to tolerate. One airline official said, in private, that "our company's

33 "Sir, do you think we're going to get out of here?"

financial position is not affected if we have one major crash each year"!

In most countries (including Britain, but excluding the United States, Australia and Japan) the law does not require airlines to report incidents other than those in which death or injury was caused or where the aircraft was substantially damaged. But if we look at the records of a country like Australia, which insists on all incidents being reported, we find that there are *ten times as many* potentially dangerous incidents as there are accidents actually causing death, injury or serious damage.[11] Since Australian airlines put up a better-than-average safety performance it can be assumed that the number of "incidents" is *at least* this proportion for all the world's airlines.

To put it another way, we could say that in ninety percent of all these incidents, the pilot and crew were able to cope; or that the trouble was so peripheral that the aircraft was not really endangered. But safety margins are constantly being squeezed tighter as the air lanes become more crowded and, as we rush pell-mell into the age of the Jumbo jets, the number of passengers at risk in any one "incident" becomes forbiddingly greater.

The pilots themselves are uneasy. This is what one experienced pilot, Captain Adrian Ross, wrote in *Flight* magazine (August 31, 1967) about British airlines:

> The constant indoctrination of the public with the perfect safety image of the national airlines, and the somewhat dubious record of the independent operators, is carried out with the cunning of a well-planned political campaign. . . . It is probably just as well that the corporations do not

[11] The committee set up by the British Board of Trade in 1967 to examine the safety of airline operations accepted this figure as probably applicable to Britain too. Board of Trade Special Review (H.M. Stationery Office).

cleanse themselves in a public confessional, as some of the incident reports, published for internal reference only, would make the average traveller wonder how an aeroplane ever got from A to B.

This is readily confirmed. Both B.O.A.C. and B.E.A. have recently been involved in pilot "work to rules" disputes wherein the captains refuse, among other things, to fly a plane with even the most minute fault. The public only has to look at the chaos of flight schedules during these protests.

But it is not only the airlines who make the pilot's job unnecessarily onerous. For example, I quote from "Judicial Investigation of Aviation Disasters," by James T. Craig:[12]

The inquiry was about a Viscount which, at London Airport in minimum visibility, lined itself up on and tried to take off from a runway no longer in use but parallel to a new runway that had come into use.[13] The system of runway marking, a code of white painted guide-marks, was provided for in the Air Navigation Order 1954, which stipulated that 'white crosses shall be displayed at each extremity of a runway which becomes unfit for use.'[14] The court accepted that there had been a white cross painted on the concrete at the very position at which the disaster aircraft had waited for and received permission to take off,[15] but it had become obliterated and had not been renewed. The court did not examine whether maintenance of the painted white cross would have contributed to safety and helped to prevent an aircraft from positioning itself in fog in entirely the wrong place and obtaining clearance to take off from that wrong position, instead it performed mental gymnastics to demonstrate that the regulations did not impose any obligation to show a white cross on the old runway. The essence of its

[12] Reprinted from *Public Law*, Autumn 1968.
[13] CAP 130—Report of the Court Investigation into the Accident of Viscount G-AMOK at London Airport on the 16th January 1955.
[14] Schedule II, Paragraph 51.
[15] By radio. The aircraft was not visible to the controller who authorized the take off.

35   "Sir, do you think we're going to get out of here?"

argument was that the old runway, 'although it had once been a runway and in some respects still looked like a runway was not at the material time a "runway" within the meaning of the rule.' So because it was not used as a runway, it was not in fact a runway, and therefore there was no obligation to show that it could not be used as a runway.

Even if this can be accepted as relevant to the court's function it can still be argued that if the old runway was not in fact a runway, then it must have been a taxi-way and it ought to have had the internationally recommended markings of a taxi-way, that is, a continuous white line down the middle which would have warned the pilot as effectively as a white cross that he was not on the runway in use. But there were no such markings. London Airport at that time did not conform to these or indeed to several other I.C.A.O.-recommended standards, but those that might have suggested room for improvement in the organisation and administration of the airfield were disregarded or dismissed. The court found that the accident was 'due to the default' of the Captain, and that neither the Ministry nor the airport authorities could 'be held responsible.'

A glance at the list of accidents to aircraft published daily by Lloyds' of London (they insure airline operators in many countries) quickly demolishes the credibility of the "perfect safety image." From this list one can calculate, taking all airports into account, that on an average three or four planes a month suffer a collapse of the undercarriage on landing and takeoff. This is one of the more dangerous incidents, because of the proximity of the fuel lines and tanks to the landing gear and the additional possibility of the fuselage sliding into some airport "obstruction."

The mechanical efficiency of the aircraft is obviously the most basic thing in air safety. Lapses in engineering maintenance occur regularly in the official accident reports as at least contributory causes of a disaster. A study of airline movements in Scotland in 1967[16] analyzed air traffic at three

---
[16] *Sunday Times*, Nov. 26, 1967.

airports—Edinburgh, Glasgow and Prestwick. Since it included the planes of a number of international airlines and covered a period of one week with average weather conditions, it can be fairly assumed that much the same maintenance and design standards would be found at any other group of airports in the world. Of the 250 flights in the week, the survey revealed, about one sixth suffered either delay or cancellation or trouble while airborne and under airport control, owing to mechanical or electrical faults. The defects included major engine troubles, even in relatively new aircraft, a host of breakdowns of electrical gear, and radio faults. In one incident all the radio receivers in a Vanguard failed when it was poised for takeoff. If it had become airborne and the ground controllers had wanted to communicate some urgent message—perhaps information about the position of other aircraft in the area—a dangerous situation might have developed.

The survey found defects in instrument-landing systems, autopilots, de-icing equipment, cabin-temperature controls, nose wheels, radio compasses, throttle controls, elevator-control indicators, oil-pressure gauges, generators, and weather radar. One VC 10 about to land at Prestwick found its undercarriage doors jammed so that it could not lower the landing gear. It was preparing to jettison fuel for a belly-landing, when the fault righted itself.

This, I repeat, is a picture of one normal week. It could not be said with certainty that any one of these defects—except perhaps the last—would have led to a major disaster. But each of them narrowed the margin of safety, if only by a hair's breadth, for the aircraft concerned. In each case, the pilots and the crews managed to cope with the faults and in most cases the passengers were left undisturbed.

In July, 1970, I was on a Trident flight from London to Geneva and, as usual, pushed my way (with permission, of course) into the cockpit. It did not take too long to notice the faults, confirmed by the pilot. Three out of the five altimeters on this aircraft were nonoperable. Two of them the automatic landing radio altimeters, and the third, the first officer's pressure altimeter.

But the fact that such a high defect rate is tolerated is a symptom of airline complacency about safety margins. To adopt the attitude that any aircraft, like any motorcar, will develop faults—faults which the pilot is capable of handling—is a failure of responsibility. The few airlines that consistently head the safety list[17] demonstrate that a very high standard of engineering maintenance *can* be achieved, if only enough care and effort are put into it all the way through the company. If these few can do it, why not the others? The backsliders include some of the biggest and the richest of the airlines.

Hard experience has shown that we cannot afford to let the safety margins get any narrower. Even an apparently trivial mechanical defect adds to the pilot's problems. If one studies the accident reports, one can see that, time and again, an accident was probably caused not by a single major mistake or mechanical fault, but by a cluster of small ones. Separately, they did not matter very much. But together they were disastrous. The pilot is tired; the altimeter is a little off; the elevator control is rather sluggish, because the maintenance inspector thought it was not worth making a fuss about. The plane runs into turbulence over a tricky terrain and is suddenly in serious trouble. The board of inquiry will not be able to pin down any one cause for the accident. It will simply talk about "contributory factors." And those are the very things that the airlines still refuse to take seriously enough.

Boards of inquiry do not examine the *general* safety standards in aviation today. So long as a plane had standard

---

[17] Refer to page 139.

equipment in working order the inquiry will accept that the plane could have been flown safely. It is well-nigh impossible after the crash to determine whether some of the equipment was in hundred-percent working order beforehand. The board seldom pauses to ask if the pilot, given the equipment he was using, had a sufficient safety margin. In too many cases the man on the flight deck is the scapegoat; he must have made a mistake somewhere along the line. Not only is this bad for the pilot, it also means that the uncomfortable facts about airline operations have been glossed over again; that we have missed yet another chance to get at the truth about air safety.

In these more general terms, the frequent use of "pilot error" in the accident reports becomes highly questionable. Of course, there are instances where the captain's judgment was badly wrong. But even here "pilot error" may be at least partly attributable to inadequate governmental control over systems for licensing air crews. However, many airlines find it hard to keep track of any one pilot's flying history, particularly before he joined the company. The result may be that an inferior pilot occasionally will get an airline job which he does not deserve, and maybe the only job he can get is with a charter airline.

Here, for example, is an extract from an Aircraft Accident Report involving United Air Lines, the biggest scheduled carrier in the Western world:

After some difficulty in acquiring the proficiency necessary to pass a practical oral, Captain Kehmeier finally did attempt his oral exam and failed it completely. He was then removed from further flight training until such time as he was able to complete the oral exam. This entailed a considerable amount of additional ground school training and took approximately three weeks. Upon satisfactory completion of the oral exam, his flight training was resumed with Flight Instructor [B]. When the areas of flight training involving the more complex aspects of pilot technique, judgment, etc., were encountered, Captain Kehmeier's performance deteriorated to the unsatisfactory stage. After approximately seven hours of instruction Instructor [B] was unable to correct the deficiencies and a Flight Manager of Standards observer was requested for the flight on February 3. Captain [C] acted as observer on this flight and his evaluation and recommendation on the basis of his observation is attached.

A review of Captain Kehmeier's record still indicates unsatisfactory performance in the areas of command, judgment, Standard Operating Procedures, landing techniques and smoothness and coordination. On the basis of the above I recommend Captain Kehmeier's DC-8 transition training to be terminated.[18]

But even skillful and experienced pilots have made "errors." It is very hard for a board of inquiry to decide how far the misjudgment was truly a personal one and how far the pilot was drawn into error by inadequate equipment.

For instance, let us take one of the simplest pieces of equipment on the plane—the windshield. It is surely not unreasonable for the pilot to insist that he should have a clear view at all times. In fact, in a heavy tropical rainstorm, his difficulties are much the same as those you would have in your car. The windshield of a big jet has a wiper mechanism which, between strokes, leaves a veil of water which reduces and distorts vision from one moment to the next. While the motorist will be driving along, for safety's sake, at a steady 20 m.p.h. the jet pilot cannot fall below his critical prestall speed of, say, around 150 m.p.h. If he happens to go off the runway while landing and hits an unseen obstruction, could we fairly describe this as "pilot error"? Or is the industry to blame for failing to solve the problem of bad-weather landings?

[18] Aircraft Accident Report, United Air Lines, Inc., Boeing 727, Salt Lake City, Utah, November 11, 1965.

2 · The Man in the Cockpit

"This is your captain speaking. . . ." The reassuring voice from the flight deck breaks in, as we cruise high over the ocean, to give us details about the weather at our destination. The message ends, and we return to our book or doze again. And this is about as close as the average passenger will get to making the acquaintance of the man who for the duration of the trip is primarily responsible for his life. Let us look at Captain Smith, who is taking out Boeing 707 "Charlie Whiskey" on a London–New York flight at noon.[1]

At about 9:30 A.M. Captain Smith arrives in his crew room at London Airport. There is plenty of work for him to do before he steps into the cockpit. His airline will have lodged an intention-of-flight document with the airport controllers, and they in turn, taking account of other aircraft departing at about the same time, will have worked out a timetable for the aircraft's departure, which allows time for every process, from taxiing out to entering the Atlantic air lanes. Once the plane is over the Atlantic, the controllers must ensure that it has enough time-and-space distance from others coming or going. The calculations are mostly done by computer nowadays, yet there still are reports of near-misses.

In the crew room the captain is joined by his co-pilot, navigator, flight engineer, and chief steward for a briefing. The chief steward reports that ninety-five passengers are coming aboard. The airline's traffic manager says that there

---

[1] Both "Captain Smith" and the call sign "Charlie Whiskey" are imaginary and do not refer to a person, living or dead, or a call sign in use at present or that has been in use in the past.

is a final tally of seventeen thousand kilos of freight to be carried if possible.

Captain Smith, not having flown Charlie Whiskey for some time (even though he has been continuously on other 707s in the fleet), inspects the aircraft's logbooks, where he will find maintenance reports and other pilots' comments. Every jet has a few special characteristics of its own. For example, it is important for the pilot to be aware of the fact that the nose landing gear in Charley Whiskey has given difficulty in the past.

While the flight-deck crew go out to Charlie Whiskey to see that things are in order, the stewards are checking the cabin arrangements and the delivery of food.

With his flight-deck colleagues the captain goes for briefing on the weather situation. He notes that there is thunderstorm activity far out over the Atlantic. This is plotted on the flight map so that he can keep careful track of it. Thinking three or four hours ahead to the time when he will be at this point, he bears in mind that he may have to ask for control clearance to move into other flight lanes to dodge this storm. He will keep a check on its whereabouts by talking to weather ships and eastbound aircraft along the route.

By 11:30 A.M. Captain Smith is lodging his flight plan in the flight-control office. This gives details of the loading and the estimated time of arrival of the flight. As the rules require, the captain nominates an alternative airport—Boston —as his landing place, in case it proves impossible to land at Kennedy Airport. He hopes to fly at 35,000 feet, known as "flight level 350."

Now the last passengers are checking in at the desk. The clerk not only is weighing the baggage but also is entering an estimated weight for each passenger. The airline manager can then calculate a weigh-bill and determine the all-up weight of the plane, including the fuel loading. He must be careful that this sum is correct. To be overweight for the runway length available could mean trouble at takeoff or a reduction in payload. To load insufficient fuel could mean difficulty if, for instance, the plane has to be diverted because of fog, or if it runs into headwinds in flight.

The co-pilot and the engineer are going through all the preflight checks of instruments according to an unvarying ritual. While they are doing this, Captain Smith is slowly walking around his aircraft, taking the same route every time. Though the maintenance men may have been busy on the plane, it is still his responsibility to carry out this tour himself, or assign the duty to the first officer or the engineer. It is imperative to ensure that control surfaces are free, that the tires are properly inflated, that there has been no fuselage damage caused by birds and so on. Finally, Captain Smith takes his seat in the cockpit for the preflight checks.

If he is a fairly old hand at flying, he will have had experience of propeller aircraft. Compared with them, the modern jet has advantages and disadvantages. It gives a vastly improved performance at higher altitudes and higher speeds. Because of its great speed it has "cleaner" streamlined wings to counteract the *drag* of the airstream. This also means that its wings give it less *lift* at lower speeds. So the aircraft has to maintain a much higher speed for a safe takeoff or landing. In order to maintain a reasonable landing or takeoff speed, the aeronautical designers have come up with *high-lift devices*—leading-edge slats, for instance, and vaned flaps, which permit the plane to become airborne at a medium "compromise" speed, and help to slow the plane down before landing.

At takeoff and at landing the pilot must keep a careful check on his speed and the plane's attitude in the air (or *angle of attack*) to avoid any risk of stalling; this occurs when a plane is not going fast enough for the flying characteristics, or *lift* generated by the wings' surfaces, to overcome the pull of gravity. The pilot must remember, along with all the other technical data, that the T-tailed jet's speed, under certain flight *empennage* patterns, takes at least ten seconds to react to an increase of throttle. The jet also loses height much more quickly than a prop-driven aircraft when speed is reduced during landing. For these reasons, the pilot needs to concentrate hard when the aircraft is in the airport circuit. When it comes to emergency decisions such as whether to *overshoot* on a landing run, or to *abort* (give up) on a takeoff run, his reaction must be very speedy. He usually has no longer than four seconds to make the decision and take the appropriate initial action.

The pilot of a four-engined jet is handling a creature of extraordinary power—far more powerful than most passengers realize when they nervously compare the tiny size of the jet pods with the great bulk of the fuselage. In fact these engines develop such thrust that the big jets can carry a payload of passengers and cargo which is nearly as great as their own unladen weight. Again, with normal payload, the aircraft can be safely flown and landed on two of its four jet engines.

But now Charlie Whiskey is nearly ready. In the few minutes from just before takeoff to the moment when the aircraft reaches 1,500 feet comes one of the pilot's most preoccupied times. He has to make calculations about a number of things: his planned weight in relation to known atmospheric conditions; his fuel consumption; what speed and rate of climb he should initiate; and the airport noise-abatement procedures.

He mentally goes through his emergency procedures, so as to be ready for the unexpected. There is his "abort takeoff" drill, which requires instant cooperation of him, the copilot and the flight engineer. There is the emergency-descent drill against the contingency of trouble when he is airborne. The instrumentation of a jet is extremely complex. The control panels around the captain will have roughly fifty separate dials, switches, knobs and levers. In an emergency, safety depends a good deal on the crew's teamwork and the mastery of all safety drills.

Recently a B.O.A.C. Boeing 707 got into serious trouble after takeoff from London Airport. The plane was only 1,700 feet up when an engine suffered a turbine-disc disintegration and fell off. At the same time, the wing and port-side fuselage were engulfed in flames. The alert ground controller warned all other aircraft to keep clear and almost in the same breath allotted the stricken aircraft a landing runway. Captain Charles Taylor brought the plane round and landed it. Five people died, eighty-three survived unhurt, and thirty-eight were injured in escaping from the wreck. Unfortunately, in this case one thing that went wrong was that someone forgot to close the fuel shut-off valve.

The pilot can never afford to let his self-discipline slacken. His routines must be so ingrained that they become almost automatic, like shifting gear in a car.

So Charlie Whiskey taxis out to the runway, the captain and flight engineer carefully monitoring the engines for any signs of faults—surging, uneven revving, overheating, and the like. As he awaits permission from the control tower to take off, the pilot will bear in mind the temperature of the tires and brakes during taxiing and takeoff. On a jet like the Boeing 707 each of the eight wheels on the main landing gear carries anything up to seventeen tons. The weight of the aircraft, no matter what type it is, induces heat; so the pilot must go gently and use his brakes sparingly during the taxi run. He knows that if he has to abort the takeoff after reaching "unstick" speed, the temperature of the brakes and tires could reach 200 degrees centigrade risking wheel lock and unsteerability. In very hot climates, where this danger is obviously enhanced, a pilot forced to return to an airport soon after takeoff will usually, if the emergency is not too pressing, cruise around for a while with the wheels down to cool them off, while jettisoning fuel to reduce landing weight. Fortunately, these days a large number of planes have automatically controlled braking equipment that minimizes these risks.

A takeoff in wet weather needs extra care, and the pilot's flight manual makes it clear that more than a quarter of an inch of slush or water on the runway can be very dangerous. One reason is that water is bound to be sucked into the engine pods as the plane gathers speed. This water is usually thrown up by the nose wheel and although engines can vaporize a great deal of it safely, beyond a certain point the jet flame will be extinguished and the pilot will have lost the power of one or more engines at a crucial moment. A more important problem is *slush drag*, requiring a longer ground run before taking to the air.

There is also a specific braking problem on a wet surface. Not only does it take longer to lose speed in water and slush but the pilot's ability to control the plane may be greatly reduced because of *aquaplaning*. This occurs when the wheels ride on a sheet of water as though it were a sheet of glass and lose contact with the ground completely. With skill these conditions can be controlled by using ailerons and rudder as well as air brakes.

Charlie Whiskey, now two thirds down the runway, has reached V-1 speed, the speed at which the pilot must decide "go or no go." To ensure a safe takeoff he waits until the co-pilot calls "Unstick" or "V R." The few extra moments give the plane extra speed, but also mean that it has reached the point where the pilot must get it airborne. At V-1 there is just enough runway for the pilot to start braking to abort the takeoff, but a few seconds after V-1 is called he cannot turn back; he has to fly.

After climbing to about one thousand feet, the pilot throttles back, and the passengers will hear a change of engine note. This is purely to meet the airport's noise-abatement regulations. If the sound-monitoring apparatus on the ground records excessive noise, then the pilot will be reprimanded. So he reduces engine speed and his angle of climb for a while, then applies full power again until he reaches his cruising altitude of 35,000 feet. Then, having checked that everything is in order, the captain switches on fully the automatic pilot, which will maintain the speed, height and compass course set. Many pilots go "partially automatic" soon after a 2,000-feet height is reached.

"Going automatic" naturally does not mean that the man in the pilot's seat can relax (the co-pilot will, of course, be taking his turn there). First, he must watch for other aircraft. The reader might suppose that this is an easy job, considering the vast air space over the Atlantic and the fact that each aircraft is carefully routed between Europe and North America in its own flight lane. In fact, the sky is not quite as wide as it may seem. The demands of economical flight routing between airports and the fact that jets operate best only within certain height limits mean that traffic corridors are not that far apart.

The safety authorities have tried to ensure adequate separation by fixing flight levels in "several layers" with 1,000 or 2,000 feet of empty space between them. Laterally there are a relatively small number of corridors, 120 miles apart. But the big increase in trans-Atlantic traffic means that at certain times of day there is a heavy concentration of aircraft in the eastern or the western part of the ocean. This system would be absolutely foolproof if flying were always perfectly accurate. Of course, it is not. Jets on the Atlantic run (cruising at nearly 600 m.p.h.) have been known to stray ten or twenty miles off their course and in an occasional case, as much as fifty miles, due to navigational errors in strong cross winds. It needs only two or more planes in adjacent corridors and making serious positional errors at the same time, for the safety margin to be narrowed sharply. There is also the possibility that planes supposed to be flying at different altitudes may collide. This possibility exists because no high-altitude radio altimeter has yet been devised, and some barometric altimeters have been proved to give false readings under certain atmospheric conditions.

One of the strangest aspects of this part of the aviation story was the effort made by the airline companies and governmental bodies in 1966 to have the space between trans-Atlantic air lanes reduced from 120 miles to 90, so that more traffic could be squeezed in. Fortunately for the passenger, this move was blocked by the vigorous and determined opposition of the International Federation of Air Line Pilots Associations which pointed out the risks. This episode reveals, in the starkest possible way, the airlines' willingness to narrow safety margins for the sake of extra revenue.

And so the pilot, even over the Atlantic, keeps a lookout for other traffic. He will also constantly monitor the engines, the cabin pressurization, and give close attention to the weather ahead. The weather radar apparatus in Charlie Whiskey will, by now, have detected the core of the thunderstorm 150 miles ahead. Captain Smith requests permission from the ground control to move up to 38,000 feet, where he thinks he will be above the main disturbance.

There will be a number of weather hazards that any pilot will try to avoid. In the freak atmospheric conditions created by a storm there is a possibility that navigational equipment will not function reliably. Structural damage may be caused by cumulo-nimbus turbulence or by hailstorms.[2] Hail may also cause a "flame out" in one or more of the jet pods, if the engines ingest too much of it. Fortunately, hailstorms are

[2] In 1969, a VC 10 flying from Bahrein to London had its nose cone smashed off by hailstones: the nose cone flew up and struck and damaged the T-tail. The pilot was able to land safely.

usually short-lived. If the pilot is caught, the quickest way out of trouble is to reduce speed and keep going. Ice is no real problem to the modern airplane so long as the pilot keeps an alert eye for any sign of a buildup along the wings, tail and fin, or around the engine intakes, or in the engines themselves. There are heaters built into these which can be armed to melt the ice. Lightning is another lesser worry; in the last twenty-two years there have been only two accidents solely attributable to this factor, although there have been many instances of large holes being burned in the fuselage.

The most serious weather hazard, clear-air turbulence, is the one about which least is known. It has only been discovered with the coming of the modern, high-flying, high-speed jets, and is associated with fast-moving airstreams at high altitudes. It can exert tremendous buffeting forces on a fuselage. It has already been responsible for a number of disasters to apparently sound aircraft. But it is a subject that can best be dealt with later, when we examine how well or how badly the airlines equip their pilots to meet the unforeseen hazards of this kind.

Having flown safely around the storm area, the pilot has told control at Kennedy Airport that he will be landing there in about one hour. It is at about this time that he has to receive permission to begin his routine maneuvering to lose height, then speed. He also has to monitor his instruments carefully to ensure that he is not losing either one too quickly. For a trained pilot this is neither more nor less difficult than most of the other judgments he will have to make in the course of the flight. Still, he must lose speed and at the same time prevent a stall situation.

Some T-tailed aircraft are, or were, capable of *deep stalling;* this means, in simplified terms, being in a stalled condition without being able to bring the nose of the aircraft down. It is practically impossible to recover control once the plane has entered a deep stall.[3] It is to guard against this danger that the apparatus called the *stall warning siren* has been devised. As its name suggests, the siren sounds when a plane is approaching a stall condition. It has been calculated that the siren sounds on about one percent of all flights; or, to put it in a more graphic way, if you have made a hundred flights, you, the reader, will statistically have been in this near-stall position at least once.[4] The pilot will take immediate corrective action at this point, either increasing engine speed or putting the nose down, or more probably both if he has the altitude to permit it.

As a safeguard there is an automatic stall corrector which comes into operation on about one in every thousand flights, when the pilot's correction has been insufficient. It is called a "stick pusher." The plane actually enters the stalled condition on about one in every hundred thousand flights.[5] Even when this happens, the pilot can get the plane out of the stall, but it is a highly dangerous maneuver, especially near the ground. In tests an unladen T-tailed aircraft has lost 24,000 feet in altitude before recovery from a near stall was effected.

The controller at Kennedy gives permission for Charlie Whiskey to continue its approach down to 14,500 feet, where the pilot must go into circuit in the "north stack" and await instructions. This is again a period of careful maneuvering for the pilot. Given the go-ahead to land by control, the pilot adjusts his speed and descent angle to enter the correct glide slope. This ensures that he will be about nine hundred feet up when he is three miles away from the edge of the runway. He must be most accurate with the setting of his throttles. Some planes, especially T-tailed craft, respond

[3] *Handling the Big Jets*, R. B. Davies, A.R.B. Publication, 1967.
[4] *Ibid.*
[5] *Ibid.*

rather slowly to upward changes of thrust when prepared for a landing, and the pilot wants to handle the throttles as little as possible during the landing, because of the risk of *underspooling* and stalling, on the one hand, or overthrottling and burning up the runway on the other.

He must keep a close eye on the effects of any wind and make corrections for it. While the over-all strength of the wind decreases as the aircraft loses height, it becomes more gusty and irregular because of the buildings and other ground obstructions. So his eye must flicker constantly up and down from the approaching runway to the speed, height, and glide-slope instruments on his panel. This essential procedure can cause misreadings very easily when he is working with clock-face–type instruments.

Then comes the actual landing. The main undercarriage wheels must touch down as close to the threshold as possible so that the plane is given more than adequate space in which to stop. At those airports where the runway length is minimal (economy beating the safety margin again), or in wet weather anywhere, this can be vital. Because he is seated some sixty feet forward and ten feet above the main landing wheels, the pilot must judge very carefully the moment when he brings up the nose to let the main undercarriage wheels touch the concrete first. In that moment (the "flare-out"), he cannot see the runway ahead, so he must take care that he has come far enough along the strip to miss the softer ground just before the threshold. The normally accepted "wheels touch" distance is between a thousand and fifteen hundred feet down the runway.

As the air lanes become more crowded and safety margins that much narrower the airline pilots have persistently appealed for improved equipment.[6] This in itself should be enough to stir public curiosity about the true state of aviation safety; but the "indoctrination of the public with the

[6] Air Line Pilots Associations Annual Air Safety Forums.

perfect safety image"—to quote Captain Adrian Ross mentioned in Chapter 1—makes these warnings ineffective.

Take the altimeter, for example, possibly the most vital instrument on the plane. Most passengers would agree with the pilots that it should be good enough to offer the captain an accurate and crystal-clear reading of his height above ground at all times. What do we find? All passenger aircraft are equipped with a barometric altimeter which reads height by measuring air pressure. Three types of error occur in using them, and although they have been several times exposed by accident inquiries, little has been done by the airlines, the instrument manufacturers or the design researchers to correct them.

First, the barometric altimeter needs to be manually set (from information passed by radio) to agree with the ground air pressure prior to a landing. An incorrect setting through mishearing or other reason can mean a difference of hundreds of feet in a pilot's calculations for landing.

Second, under certain abrupt changes in barometric pressure the reading is likely to be false for a short while. If a plane is trying to locate a runway in fog and descends to, say, thirty feet and then suddenly rises to, say, sixty feet, the pressure altimeter could give a wild reading for a few moments. This has happened and has been cited as a probable cause of a major disaster. At sixty feet there is no margin for misinformation of this kind.

Third, reading errors of an almost unbelievable magnitude are possible because of the design of some altimeter dials. The main barometric types have one, two or three needles and a drum. Each unit represents tens of thousands, thousands and hundreds of feet above the ground. It is possible with the three-needle type, for one needle to obscure another or for the longer needle to be mistaken for one of the others. It is, of course, one of the most dangerous errors in instrument reading, as several disasters have shown.

For example, in December 1958, the pilot of a B.O.A.C. Britannia, on a daytime test flight over southern England, requested permission to descend from 12,000 to 3,000 feet. Three minutes later, the aircraft struck the ground, which had been obscured by fog. Nine of the twelve on board were killed. The subsequent inquiry[7] said it would seem from the evidence that the Britannia was not at 12,000 feet when it started its let-down, but at 2,000 feet.

The inquiry recalled a similar accident, which had taken place earlier that year[8]—another 10,000-foot error, which was probably due in part to needle misreading. On that occasion, the Board of Inquiry had urged pilots and co-pilots to cross-check their altimeters to avoid this mistake. It added:

> The height presentation afforded by the type of three pointer altimeter fitted to the subject aircraft was such that a higher degree of attention was required to interpret it accurately than is desirable in so vital an instrument.

The board, as I have said, saw fit to advise flight crews on the need for cross-checking. Surprisingly, it did not come out with really vigorous criticism about the risks involved in employing such a confusing instrument at all.

This was by no means the last accident in which the altimeter was held to have been the main culprit. A Viscount of Austrian Airlines which crashed on approaching Moscow Airport in September 1960, with a loss of thirty-one lives, was, according to the Board of Inquiry, probably the victim of some form of altimeter error.[9] A Finnair DC 3 which was destroyed (with twenty-two people killed) when it hit trees on a landing approach in November 1963, had probably, according to the Board of Inquiry, been flying on a wrong

---

[7] I.C.A.O. Circular 62 AN/47, p. 44; I.C.A.O. Accident Digest No. 11.
[8] British European Airways—Viscount—Apr. 28, 1958—I.C.A.O. Accident Digest No. 10, p. 129.
[9] I.C.A.O. Accident Digest No. 12, p. 266.

altimeter indication.[10] Misreading of the pointers by ten thousand feet was probably, according to the Board of Inquiry, a contributory factor in the loss of a United Air Lines Boeing 727 in Lake Michigan in August 1965.[11] And the Iberia Air Lines Caravelle which flew into a low hill in southern England in November 1967 (with the loss of thirty-seven lives) at a point on his route where the pilot must have believed the plane to be at 10,000 feet, also points to a probably drastic error somewhere in the altimetric system, or in the reading of the instrument.

Quite often a height error is combined with other factors in causing an accident. But this can only mean that altimeters are misread or wrongly set even more often than the accident reports suggest; and the mistake is simply waiting for some other difficulty, like bad weather, to turn it into a tragedy. There was, for example, the Air India Boeing 707 which crashed into Mount Blanc in January 1966, with the loss of one hundred and seventeen lives. The pilot, according to the official French inquiry, probably "incorrectly estimated" the height at which he was flying. The Geneva radar controller had pointed out in French to the pilot that he was flying too low, but the inquiry report says that probably the information was not clearly understood.[12] (English is regarded as the common language of aviation wherever the plane is flying and regardless of the nationalities of the flight crew and ground control.) Given better height instrumentation, the plane need never have been at risk at all.

There have been similar accidents during the 1960s. Yet the barometric altimeter of the three-needle type is still in widespread use as an aircraft's sole instrumental guide to height. Until someone comes up with a new, efficient way of evaluating height, a dual system of altitude measurement is

[10] I.C.A.O. Accident Digest No. 15, Vol. 11, p. 201.
[11] N.T.S.B. Report, *Flight International*, Feb. 1, 1968.
[12] French accident report, *Flight International*, Mar. 21, 1968.

clearly needed, making use of a radio altimeter as well as a barometric one. No more than one in twenty of the world's airlines have radio altimeters installed in some of their planes, despite proof of its great value to the pilot, especially in difficult conditions of weather and visibility. It works by "bouncing" a radio signal off the ground and converting into height the time taken for it to return to the aircraft. It has a digital information display, which is easier to read than the clock-face type. One of its shortcomings is that it is reliable only up to 2,500 feet. Even so, it could be a check against any misreading of the barometric type at low levels, where accuracy is most needed.

A horn or siren can easily be connected to the radio altimeter circuitry to attract the pilot's attention to the fact that he has descended within 2,500 feet of the ground. The radio altimeter also has an uncomplicated dial. It is invaluable on landing approaches because, at these lower levels, the pilot can get an accurate reading to within a couple of inches. And equally important, there is no manual resetting to be done before a figure can be obtained from it.

Why, then, do we not find a dual system fitted on *all* our big jet liners? The short answer is that it would cost five times as much money as the barometric instrument alone.[13] The airlines evidently think that this is too high a price to pay for greater safety, despite the number of height-error accidents over the years.

Only now are the dual radio-barometric altimeters being installed in more aircraft because this system is essential for automatic landings. How many planes have automatic-landing systems? How many planes that are equipped with automatic-landing systems are using them? The answer to both questions is: Practically none.

Making matters easier for the pilot is no longer merely desirable but essential. The pilots, for instance, have for a

[13] Estimate from a United Kingdom altimeter manufacturer.

long while been asking for some form of pictorial navigational display that will instantly show them their position in relation to the flight's progress.[14] It would relieve the understandable anxiety that the pilot feels when he is putting his trust in the radio location signals from the ground en route (the traffic controllers' radar) or is "flying blind," with the ever present possibility that there will be a breakdown, leaving him suspended, suddenly not knowing his position or sighting.

The Senior Accident Investigator for the British Board of Trade, Captain V. A. M. Hunt, writing in the British journal *Flight Safety Focus* in November 1967, put forward evidence to confirm the need for this sort of equipment. Discussing collisions with high ground, he commented that there was "mute evidence" of the "false sense of security that can be engendered in the air." He added:

> My air safety colleagues, during the course of an analysis of world jet approach and landing accidents, have come to the conclusion that, out of twenty-three which they analysed in detail, some fourteen might well have been avoided if the pilot had known precisely where he was, particularly in relation to the ground. When executing what can often be complex manoeuvres during the approach to land, this facility can only be provided continuously by pictorial display, driven by appropriate navigational aids.

But, again, the airlines and the airport authorities fight shy of the extra cost of this equipment in the planes. Governments cannot spare the costs of ground-to-air "nav-aids." So they leave the pilot with his problem.

Mid-air collisions remain a haunting problem for all concerned with aviation. At the speed of today's jets, what looks like a harmless situation, such as sighting another plane a mile or two away, can turn into a dangerous encounter be-

[14] British, American and International Federation airline pilots' associations all have fully discussed the need for such equipment.

fore either pilot has had time to react. In fact, under the very best of sighting conditions, it has been found, a pilot can see another aircraft on a head-on course at a maximum distance of only seven miles, which, with a closing speed of around twelve hundred miles per hour, leaves only a few seconds for avoidance action. A lengthy study of collision accidents has shown that about four out of five took place in daylight, when conditions were good enough to permit a sighting of the plane.[15] Most of them happened within twenty miles of an airport. In the last twelve years five major mid-air collisions in the United States alone have taken a toll of 449 lives.

One of the worst so far was the collision between a Lockheed Constellation of Trans World Airlines and a DC 7 of United Air Lines over the Grand Canyon, in Arizona, in June 1956. Both aircraft were destroyed, and there were no survivors among the one hundred and twenty-eight persons on board them. The official inquiry simply concluded that one of the probable causes of the accident was that the "pilots did not see each other in time."[16] Its report did not fully discuss the questions naturally raised in the mind of the average passenger, such as why aircraft are allowed by the A.T.C. to fly with such limited height separation on crossing paths, when the enormous sighting difficulties are so obvious and when collision-avoidance equipment is so manifestly lacking. Nor did the inquiry suggest any remedies. The report *did* mention that a contributory factor was the pilots' "preoccupation with normal cockpit duties." But it should be obvious that safety equipment on the flight deck will remain inadequate until it is accepted that the pilot is *always* to some extent involved with problems other than sky-scanning. We cannot totally rely on miracles of eyesight and concentration at these speeds.

[15] Air Line Pilots Association Air Safety Forum 1967.
[16] I.C.A.O. Accident Digest No. 8, p. 95.

The Man in the Cockpit

It has become imperative to develop and install new safety devices on airliner flight decks as well as to modernize the traditional ones, like the antiquated altimeters. A reliable collision-avoidance system is urgently needed. Air-traffic-control reports show that aircraft frequently come within dangerously close range of each other at all major airports. In the United States alone it has been calculated from official reports that the number of near-misses has risen to more than four hundred a year. A. M. Lester, until recently Director of the Air Transport Bureau of the International Civil Aviation Organization has said that in the next five years there will be about a dozen mid-air collisions caused by congestion around airports.[17] Since 1960 the number of flights around airports has gone up by about 75 percent. The increase in sheer numbers, as well as the variety and speeds of the aircraft in the traffic mixture, has made the problem of spotting other planes from the flight deck an acute one. It is necessary to make absolutely clear that anxiety about the situation does not come merely from the uninformed passengers. It is the pilots, the controllers, the aviation officials —the men who *know the facts*—who are seriously alarmed. Discussions at virtually every air-safety meeting show this beyond a doubt.

David Thomas, deputy administrator of the Federal Aviation Agency, the principal architect of the air-traffic-control system in the United States, gave his estimate of the probable increase of civil air traffic in America. He predicted that within the next ten years the number of airliners would increase from the present 2,100 aircraft to 3,500. The current 100,000 planes in the general aviation fleet (which includes civil, business and all private planes) would increase to about 180,000, with fast jets alone becoming sixteen times more numerous. All these additional aircraft are not going to spread out into new and untried air space. Be-

[17] Air Line Pilots Association Air Safety Forum 1967.

cause economical routes between airports remain the same, they will nearly all have to be fitted into the air routes already in use.

This intensifying of the problem has been predictable for some years. Yet it was not until 1967 that the Air Transport Association in the United States (which is more safety-conscious than other countries) set a group of electronics manufacturers the task of devising a reliable collision-avoidance system. The system they have suggested would involve fitting a small computer to *each plane*. The instrument would calculate changes of distance between aircraft and would send a coded radio warning to the other aircraft if any two of them got within sixty seconds' flying time of each other. If they continued to close, then at forty seconds' range an indicator would instruct the pilot whether to dive, climb or hold steady to avoid a collision. One official has predicted that this sort of equipment will be in general use in 1974. By that date it will have been pressingly necessary considering the extra volume of air traffic in circulation, with its mixture of Jumbo jets and supersonics.

Even now, however, there are many situations such as the one described in a report on the Manila–Hong Kong route:

> It was reported that the airways from the Philippines to Hong Kong (Airways A61, A83 and B72) cross airway Red 71 at almost the same position. This latter airway (R71) is one of such high traffic density that flight levels other than those according to the correct semicircular rule are being utilised.
> At the same time traffic along the Airways A61, A83 and B72 is often not known to the Hong Kong control authorities because of the poor communications link from Manila to Hong Kong.
> Under the circumstances a considerable risk of collision exists.[18]

[18] International Federation of Air Line Pilots Associations, Report of the 23rd Conference, Oslo, Norway, March 29–April 5, 1968.

The Man in the Cockpit

There are a number of ways in which a change of attitude could show improvements without any need to invest big sums in equipment or research. Here is a simple example that is particularly relevant to the mid-air collision risk. With a number of medium-size jets, like the BAC 111, the Boeing 737 and the DC 9, the airlines have claimed that a two-man flight-deck crew, the pilot and a co-pilot, is adequate.[19] This is the way some airlines now operate them, despite the protests of the pilots' associations that three men are needed in the cockpit.

The pilots point out that there have been numerous cases of at least one of the pilots becoming ill during a flight. On a number of occasions the illness has been so severe that the pilot in question has been put severely out of action. In other cases his condition would have made it foolhardy to let him continue handling the controls. Answers to a questionnaire on the subject, sent out to all American airline pilots by their association, showed that out of 2,562 captains, no fewer than 1,519 had at some time been incapacitated by illness during flight, while 863 out of 1,982 co-pilots had had the same experience. It was emphasized that the illness had had to be severe enough to make it necessary or desirable for another crew member to take over the sick man's duties.[20]

The most dramatic form of incapacitation is death in the cockpit. Between 1952 and 1966 there were sixteen cases in which a pilot employed by a United States airline died suddenly on duty; in nearly every case it was from a heart attack and there was little or no warning. More than once it happened at a critical point in the flight. Fortunately, another crew member was usually quick enough to take over and avert a disaster. Only one of these incidents resulted in a crash that the board of inquiry attributed directly to the

[19] B.A.L.P.A. Report and Air Line Pilots Association Air Safety Forum 1967.
[20] Air Line Pilots Association Air Safety Forum 1966.

incapacitation of the pilot. This was the crash of a Flying Tigers Line Lockheed Constellation during a landing operation at Burbank, California, in December 1962, when the captain died suddenly at the controls. Seven others died with him.[21]

A similar incident involved a Trans Australia Airlines DC 4, landing at Brisbane, Queensland, on May 24, 1961. The official accident report says that what probably happened was that the pilot in command suffered a heart attack during the crucial period of let-down and, in trying to evacuate his seat, must have fallen across the throttle controls, pushing them all into the closed position.[22] With this presumed sudden loss of power, the plane crashed to the ground before the other members of the cockpit crew could drag the body off the controls.

If one pilot is put out of action, then in most cases his colleague in the other cockpit seat will get the plane down smoothly and safely, though he is having to handle the radio, monitor all instruments, think of the navigational aids, keep a lookout and fly the plane singlehanded. But the safety margin has obviously been sharply cut back, and it will be another of these occasions where the shortcomings in equipment will present a double degree of risk.

The pilots have emphasized the usefulness of having a third man on the flight deck as an extra lookout, a valuable insurance against mid-air collisions. They would probably stand a better chance of discouraging the airlines from crew economies if the public were more aware of the alarming frequency of near-misses. Collisions can happen for any of a variety of reasons. Some aircraft in the airport circuit may be flying by instruments, some by visual means. Small private aircraft often involved in these incidents can be hard to spot quickly. Through radar defect or oversight, a

[21] Air Line Pilots Association Air Safety Forum 1966.
[22] I.C.A.O. Accident Digest No. 13, p. 78.

ground controller may completely miss one aircraft and thus fail to give others a warning about its presence.

As the figures show, near-misses are a more than once-daily occurrence in the United States alone.[23] Some examples of these close shaves and their frequency were recounted to the 1967 Air Safety Forum of the American Air Line Pilots Association by W. W. Betts, the central safety chairman. He spoke of the experiences of one young association member who had played a vital part in collision avoidance on three occasions—in a period of only two and a half years with the airline.

The first scare [this member had written] was in my first six months as an engineer in a Caravelle. While letting down [descending] between East Texas and Rocky Hill, with the co-pilot flying the aeroplane and watching his instruments, and the captain talking on the radio, I saw a dot directly ahead that grew in size so rapidly I had no time to try to point out the target. I immediately grabbed the wheel from the first officer, pulling back and down, which caused us to climb to the left. The captain caught a glimpse of it as the small aircraft went under the nose, but the first officer said he saw nothing . . . and we all went home just a little shaken up.

On the second occasion, breaking out of cloud over Cleveland Airport his Boeing 727 came "nose to nose" with a Navion aircraft and the pilot, hearing the third man's warning shout, pulled back the stick just in time to climb over it.

The third near-miss was the most hair-raising. The informant was flying a Boeing 727 on an approach to Kennedy Airport. "The first officer suddenly hollered, 'Two targets at 12 o'clock.' We were able to pull up slightly and fly under an Electra that was head-on with us, and just over another Boeing 727." Kennedy Control said there was nothing showing on the radar screen in the vicinity. This may really have

---

[23] Civil Aeronautics Board annual figures.

been an oversight of the approach controller, but the first officer was the one who saved it from being a very close miss.

The British Airline Pilots Association—like its counterparts in other countries—has numerous similar reports on its files. Mr. Betts has observed:

> Mr. Lester's prediction of the number of mid-air collisions to be expected, may prove to have been too conservative, considering the remarkably high number of times when the third man in our present-day jets has been the first to spot collision traffic and alert the crew member at the controls.

Instrumentation to detect special and possibly dangerous weather conditions is adequate in most commercial aircraft. Pilots have been fairly well accustomed to dealing with ordinary turbulence ever since aviation began, though it can sometimes mean an extremely rough ride for the passengers. "Normal" turbulence is usually associated with updrafts and downdrafts of air, in and near thunderstorms, cumulonimbus cloud activity, and mountainous terrain. A plane that has failed to avoid this kind of air disturbance or has to fly through it can gain or lose five thousand feet from its normal flight level (another reason for good collision-avoidance equipment), and the pilot will need all his concentration to keep control. If he is lucky, the weather forecast will have given the pilot the approximate position of a storm center so that he has ample time to set a course around or above it. But in periods of bad weather ground controllers get overworked and weather forecasts are quickly out of date. In a plane moving at nearly ten miles a minute through changing weather systems, a pilot has to think well ahead. To help him he has a weather radar unit which gives him more than ten minutes' warning of this kind of storm disturbance in his track.

But the high-altitude flying which came in with the jet era revealed the existence of that new, and still somewhat

The Man in the Cockpit

mysterious, weather hazard known as "clear air turbulence." One important difference between "CAT," as it is called, and ordinary storm turbulence is that the latter is almost always visible because of the lightning, ragged cloud or rain and ice that show up on the radar. CAT, on the other hand, is always invisible; it cannot be detected by radar, because there is no precipitation of moisture for the apparatus to register, and it is generally much more violent than ordinary turbulence.

CAT is caused by unseen jet streams of air moving at relatively high speeds. One of the most sinister aspects of CAT is the "shear" effect on aircraft flying at high speed through air which is moving first in one direction and then another. This is caused by the reciprocating gusting of the winds. A pilot may find that he is being forced downward in one of the jet streams. He may have time to adjust the controls to correct this, only to find in the next moment that he is being forced upward. It is presumed, as it is quite difficult to research CAT, that there have been occasions when the speed of the winds, the directional changes of the air currents, and the speed at which the plane is traveling through the turbulence, combine to create such fantastic forces on the fuselage of an aircraft that it literally "breaks apart." These forces naturally amplify any minor structural defect either ignored or unchecked during maintenance and overhaul. A hairline crack in the fuselage is inviting disaster in these conditions.

In the early years of high-altitude flying, the effects of CAT mystified accident investigators. Now it is known to be responsible for millions of dollars' worth of damage each year to military and civil aircraft. It has played a part in a number of crashes. In 1960 a Lockheed Electra of Northwest Orient Airlines, had its right wing torn off and dived 18,000 feet. The crash killed all sixty-three people on board. The inquiry said that one of the probable causes of the crash

64   Unsafe at Any Height

was oscillations of the No. 1 engine nacelle (that is to say, the wing was literally shaken off).[24] It was later discovered that the Electra wing construction was probably a contributory factor, but the plane's entry into an area of severe turbulence probably set off the oscillations. Following a similar accident (Braniff Airways, Sept. 29, 1959),[25] the Electra was permitted only restricted operation certificates until the Lockheed Company had made wing modifications.

In 1963 a Northwest Orient Airlines Boeing 720 crashed into the Florida Everglades. Forty-three people died. The official inquiry said that probably the plane had been caught by the "shear" effect and that it had probably wrenched apart as the pilot lost control.[26] In August 1966 a Braniff Airways BAC 111 broke apart in the air over Nebraska. This again was probably because of clear air turbulence, and all forty-two people on board were killed.[27]

In March of the same year the biggest CAT disaster occurred when a Boeing 707 of B.O.A.C. broke up in flight near Mount Fuji in Japan, with the loss of one hundred and twenty-four lives. A Japanese board of inquiry said that probably it had suddenly encountered turbulent winds of tremendous force.[28] Soon after this accident B.O.A.C. took half its remaining 707 fleet out of service because they found hairline cracks in the tail assemblies of these planes.[29]

For every case of a crash or severe structural damage caused by CAT there are believed to have been at least one hundred near-misses in which the coolness and skill of the pilot, and luck, extricated the plane from a potentially lethal situation. A Boeing 720 jet was forced to make an 8,000-foot dive during a flight over Wyoming in March 1967, when it

[24] I.C.A.O. Accident Digest, No. 12, p. 137.
[25] *Ibid.*, p. 51.
[26] *Ibid.*, No. 15, Vol. 11, p. 99.
[27] N.T.S.B. Accident Report 1-0008 SA 303.
[28] I.C.A.O. Accident Digest, No. 16, Vol. 11, p. 35.
[29] London *Times*, Apr. 21, 1966.

entered a CAT area. The pilot managed to pull out of it. One passenger was killed; he suffered a heart attack when he was thrown from his seat.[30] The strength of CAT gusts can be better assessed if it is appreciated that a jet aircraft is designed to withstand heavy stresses, so long as they are applied fairly evenly.

The instrument manufacturers have come up with an idea for a CAT detector.[31] It works on the principle that there are marked differences of temperature between one side of a CAT wind-shear and another. An infrared detector mounted outside the cabin can pick up these variations to a fraction of a degree, at a range of nearly fifty miles. It has been able to give effective warning to the flight crew, and is experimentally installed on a very limited number of planes in airline service today.[32] This is encouraging news for everyone who flies. But one cannot help reflecting, sadly if not angrily, that the time lag involved in getting this sort of development moving has been considerable and there is still no assurance that a CAT detector, or the other instruments needed, will be general equipment on jet liners this year, next year, or the year after. In any event, the device mentioned above would be of doubtful use to the supersonic airliners, since the plane would hit the turbulence within a few seconds of the device detecting it.

But development time lags are a feature of the industry's history. No effective measures have yet been taken to deal with what is probably aviation's most basic hazard—the pilot's lack of certainty about his position, on the map and in relation to the ground. The problem has been there ever since the first "string-bag" planes of the early days. If commercial aviation had given proper financial backing to re-

[30] *Time* magazine, July 7, 1967.
[31] Barnes Engineering Inc. CAT Detector.
[32] Evaluation of Infrared Spectrometer as a CAT Detector, N.A.T. Research Council of Canada.

search in this field we should not be in such a dangerous situation today. Planes fly at greatly increased speeds and with many times more passengers than was the case even fifteen years ago. Yet much of the equipment for ascertaining position is not notably better than it was in the days of the biplanes and ten-seaters.

All the experts in the different fields of civil aviation seem to be agreed that the greatest boost to safety we could have would be the mandatory introduction of a modern worldwide system of identifying high ground and other solid obstructions in the plane's path. Once off the major air routes in common use, the pilots say, there is a serious lack of ground-navigation aids. So there are still a large number of accidents in which pilots have flown into high ground totally unaware that they were on a dangerous course. It is surely reasonable to say that on the verge of the supersonic era, this kind of crash should be extremely rare, and the risk of it virtually nonexistent for big passenger aircraft; but this is not so.

The siting of ground navigational aids is the responsibility of the respective governments, and unhappily one cannot expect them to give this matter a high degree of priority. There are, however, more basic improvements they might be urged to undertake, such as making airports less dangerous.

Ideally, the pilot needs a "pictorial display"—a map display with crossed lines showing him exactly where he is and also showing him his all-around position. But there are simpler and cheaper devices which would help to take some of the most old-fashioned dangers out of flying. There is, for instance, the APQ 107 Radar Altimetric Warning System—a complicated name for an instrument that gives the simplest sort of protection. Combined with a similar radar device, it would give the pilot adequate warning, both visually on the instrument panel, and audibly by the sounding of a loud buzzer, that his height had fallen below a preset minimum

The Man in the Cockpit

or that he was on a collision course with some high ground ahead. Its cost, uninstalled, has been estimated at around one thousand dollars, which must be a great deal less than one set of in-flight movie equipment. Or, if the airlines were ready to invest in something more sophisticated, there is the type of "hedgehopper" unit already installed in the American military plane, the F-111. This is a very accurate instrument designed initially to enable the aircraft to fly safely only a few hundred feet above the ground. But it can be converted into a device which will alter the plane's course should it be flying head-on into terrain. This unit would cost about ten thousand dollars.

There are, of course, many other factors that directly affect the pilot's ability to fly his plane safely: the flying qualities of the aircraft itself; the airline's standards of maintenance on the machine; the safety of the fuel carried; the design of the airports. Each of these will be examined in later chapters.

At least one point, I hope, is clear from the evidence I have given here. When we see an aircraft disaster attributed to "pilot error," our first reaction should be to ask how far the complacency and neglect of others lies behind this easy phrase, how far the penny-pinching economies and the absence of real concern denied the man in the cockpit (and therefore his passengers) a square deal.

3 · The Overcrowded Airports

Most airports are less safe than they should be. A few are well designed and well maintained, and enable a flight crew to deal confidently with the critical period of takeoff and landing. At the other end of the scale, mostly on national domestic routes, there are some quite busy airports that are potential death traps. They strain pilots to the edge of their skill and concentration every day. Anyone who cares about the subject can only urge that these nightmarish places be closed down at the earliest opportunity.

But these extremes of good and bad must not be allowed to disguise the shortcomings of the "average" airports, those regularly used by tens of millions of passengers a year. At New York, Chicago, London, Paris, where planes move at the rate of one a minute—even faster at peak times—the most serious problem is acute overcrowding, though there are other deficiencies.

Here is an extract from a pilots' conference:

And still another burst of candor, and from an equally unlikely source, comes from the Port of New York Authority, the agency accountable to the city for the operation of its three civil airports. With remarkable courage they have noted that these three ports are saturated, that private aviation contributes forty per cent of the traffic at these ports, eight per cent of the revenue, less than three per cent of the passengers, and no cargo at all, and won't they please go some place else.[1]

[1] International Federation of Air Line Pilots Associations, Report of the 23rd Conference, Oslo, Norway, March 29–April 5, 1968.

The Overcrowded Airports

The ordinary passenger is seldom aware of the gross state of air traffic congestion at some airports. He may even have the illusion, as his plane circles for a landing, that his own aircraft is the only one in the sky for miles around. In fact, if the aircraft arrives at Kennedy Airport (New York) during the evening rush hour, when there are numerous domestic flights mixed in with international ones, it is likely to be one of seventy or eighty planes circling in "stacks" around the airport, awaiting their turn to land. Down below, as many as fifteen big jets will be queued up nose to tail waiting to be cleared to a takeoff runway. As flight crews and air controllers will agree, this is a situation that allows hardly any room for a lapse in concentration. But with this in mind, one can better understand how there comes to be over four hundred near-collisions reported each year in the United States alone (and probably a good many more unreported).

The language problem narrows safety margins still further. English is traditionally used for communication between air-traffic-control officers and all aircraft, whatever their nationality. Yet there will be a significant percentage of pilots over Kennedy at any one time for whom English is a foreign language. In the circuit there will be Frenchmen, Germans, Japanese, Scandinavians, Pakistani, and so on. They have had to learn their control-routine English in classrooms and on familiarization flights with more experienced pilots. Under normal circumstances the communications work tolerably well, but the introduction of a ground-to-air teletype system would solve the problem of A.T.C. accents. The controller's radar is a second-line check against a pilot's mishearing instructions and moving into a wrong stack or flight lane.

But consider the extra strain imposed on clear understanding in the peak periods, when the controller is calling out instructions to planes overhead in a never-ending stream. The foreign pilots must pick up the details of a message accurately, though it may be delivered fast and in a Brooklyn accent.

If the planes are going to be landed rhythmically at about one every minute, then there is barely time, the pilots say, for a captain to query an instruction or to hesitate. He must keep in mind the queue behind him. It is little wonder that even the most experienced pilots feel uneasy about such congestion. The subject is, of course, discussed regularly at the conferences of the Air Line Pilots Association, when there is pressure for anticollision instruments to be fitted on flight decks.[2] But it is worth noting that they do not take their protest to the point of rebellion. All the obvious human factors come into play. An unconscious professional pride is one of them. Unless he has had very firm orders from his airline about the conditions he is allowed to tolerate, then the average pilot will not readily admit that he cannot cope with the difficulties thrust upon him. That is one reason why the conditions which I have described at Kennedy eventually come to be regarded as "normal," though regrettable. Clearly, by any reasonable standards of aid safety, these conditions are grossly *abnormal*.

How has such an inefficient and barely tolerable situation come about? The short answer is that the industry has never, since its earliest days, anticipated the rate of technical development or foreseen the needs of the future. In 1919 there were about four small airline companies in existence in Britain, France and Holland. Few people took civil aviation seriously enough to think it was here to stay.

In those days even the short hop to Paris had to be made in easy stages, with maintenance and refueling at stops on the way. The planes of the time could land on little more than the proverbial postage stamp, a flat strip, less than

[2] Air Line Pilots Association Air Safety Forums 1966 and 1967.

The Overcrowded Airports

half a mile long. It was an era that was adventurous, sentimental, living in the moment and ignoring the future. The aviation industry took a long time to outgrow the idea that daring and danger were a natural and normal part of the aviator's experience.

By the early 1920s the first landing fields for passenger planes had been built. Then around 1935 the Douglas Aircraft Company put out their new model, the DC 3. It showed a big advance over its predecessors. It could seat twenty-one passengers, had a flying speed of over 125 miles per hour, and could land and take off in a relatively short space. Civil aviation and the DC 3 expanded hand in hand.

The airlines went on demanding bigger and faster planes with a greater flying range, and the manufacturers went on providing them. They left the problem of landing areas to local airport authorities. The development of the airplane always outstripped the development of airfields. The authorities had to find large sums of money, not only to try to keep their airports reasonably up to date in equipment but also to enlarge them—a costly business, as they need land close to city limits.

Airports are unsafe today largely because there has never been any real coordination among the three arms of aviation —the airlines, the aircraft manufacturers, and the people who run the airports. Up to, say, twenty years ago, the effects of this absence of realistic planning were not particularly disturbing, since traffic was fairly light and relatively slow.

But for the last decade and more, neglect of the problem has had increasingly severe consequences. There has been growing passenger congestion at ground level and, more important, a greater strain imposed on flight crews. The result is that congestion in airport circuits is scarcely tolerable *now* at the bigger centers, and traffic continues to mount.

Many of the major airport authorities are struggling

gamely to catch up. They have plans on the drawing boards. But once again the new aircraft will be in service before the airports are ready. Hardly any airport in the world is ready for the Jumbo jets: not only do they lack the landing and takeoff facilities but they are unable to handle passenger- and baggage-loads of this size on the ground.

It has never been demonstrated that the flying public wants to travel in planes of this size, certainly not until ample safety margins have been assured in their whole operation. Commercial pressure dictates the speed of development, a continual push to keep ahead of rivals, so that real forethought about safety is always a low priority.

Even some of the industry's own experts appear uneasy about the situation. Speaking to the Royal Aeronautical Society in London in September 1967, Dr. Richard R. Shaw, Technical Director of the International Air Transport Association, commented:

One of the greatest failures of the industry has been a persistent tendency to under-estimate its future growth and this has led to inadequate preparation in many areas—notably airports. The best forecasts of the 1960/61 era failed to estimate the 1966 actual traffic by ten to twenty-five per cent . . . I do not blame the airport authorities alone for this problem, although they must share the responsibility. The airlines themselves have generally under-estimated the growth in traffic and have not always been frank with the airport authorities on their future plans, because of the fear that their competitors might benefit from such disclosures. The manufacturers have also failed to keep airport authorities adequately informed of their forward design planning. Airport designs have tended to be too limited in scope and to be based on a traffic demand anticipated at a particular date in the future without adequate thought being given to how growth beyond that date could be accommodated.

Now, at long last, for the first time since aviation began, the sheer pressure of traffic and the rate of growth has

73 The Overcrowded Airports

forced the various branches of aviation to make a cooperative study of the problem. Will they at last be candid with each other and work on the hard facts of the situation? If so, it will not be a moment too soon. Experts like Dr. Shaw who want to see some realism in planning do not underestimate the magnitude of the job that has to be done in order to bring airports up to date.

Official estimates of the rate of growth of civil aviation are staggering enough. Putting together forecasts of the buildup of passenger and cargo traffic, it seems certain that the total traffic moved will *double* every four or five years at least for the next fifteen or twenty years.

Dr. Shaw gave some graphic examples to show precisely what this growth rate means. O'Hare Airport, Chicago, which is the largest purely civil airport in the world, can be taken as a "datum" for comparison. In 1968 it handled no fewer than 30,100,000 passengers and some 690,000 aircraft movements (landings and takeoffs).

At the expected growth rates, traffic even at an airport like Zurich, Switzerland (which might be described as "medium-busy" at the moment), with 3,500,000 passengers and 120,-000 movements, will be as heavy as Chicago's by 1975.

Even such a remote place as Melbourne, Australia [says Dr. Shaw], is likely to reach O'Hare traffic level before 1980, and Nairobi in Kenya should reach it in the 1990s. I could give hundreds of examples of this sort. It follows that every significant airport should be planned from the outset for ultimate expansion to about the O'Hare size and that any construction should be part of a long-term plan towards this end.

This is plainly a big task. It is the sort of thing that every airport authority which expects to profit from this growing traffic should be thinking about. Certainly these planned new airports are going to need enormous financial investment.

No critic should delude the public into thinking that the money will be raised quickly enough in all countries to ensure satisfactory standards in the next few years. But money is only part of the story. As I have suggested earlier it is the *attitude* to air safety standards that counts for as much.

Far too many airports have been allowed to survive into the jet age with serious and fundamental defects. These are not simply isolated landing strips in the backwoods; they are often well-known airports in populous centers, handling a large number of aircraft movements each year.[3] Indeed, sometimes the dangers are apparent even to the layman. Often the basic problem is poor siting. There may be hills all around or a lake or the ocean at the end of the runway. Planes have gotten bigger and faster; they need more air space or a better landing approach. But the airports have not developed at the same rate. Thus the burden of coping with the narrower safety margin is thrown onto the pilot. So we get accidents like the one at Cincinnati in which a Convair 990 of Trans World Airways crashed into trees 9,000 feet short of the runway on its final approach, with the loss of over sixty lives.

In some cases little can be done about the difficulty of the terrain. Hong Kong, for example, is situated on a bay with a mountain range on one side and more high peaks on islands on the other.

But at the very least it is imperative that the pilot flying in or out of Hong Kong should be provided with the very best instrumentation. Regrettably, this is not the case. The liability to error with the barometric altimeter has already been mentioned. In addition there are still far too many airports which present the special hazard of difficult terrain *and* woefully inadequate navigational aids or poor runways.

[3] I.C.A.O. Air Navigation Plan, 11th Edition; IFALPA, 23rd Conference, Oslo, 1968.

The Overcrowded Airports

Even in the United States there are a number of airports which are below standard in these respects. In Europe a few airports have earned a black reputation for the mediocre equipment and the consequent series of disasters. Some of the worst of these death-trap airports are found in the mountainous areas of southern Europe. Between 1951 and 1967, for example, no fewer than nine aircraft disasters took place in the area of Perpignan near the border between France and Spain. There are high mountains nearby and dangerously strong and shifting winds. Yet throughout this period the airport at Perpignan had very poor runway lighting, no glide-slope indicator system (an arrangement of red and white lights to show a pilot his proper descent angle in a visual approach) and not even elementary ground radar. In the case of Cairo airport, the International Federation of Air Line Pilots Associations went so far as to implement a ban on the main runway.

The problems at Cairo (International) could be attributed to the following eight points . . .

1. The influence of the environmental factors on the location of the airport;
2. Allocation of funds to satisfy current operational requirements of jet aircraft;
3. Maintenance and monitoring of existing technical facilities on a continuous basis;
4. Methods and programs for training operational personnel;
5. Non-implementation of promises already given in official correspondence with the Authorities [thus luring IFALPA into the belief that improvement was just around the corner];
6. Withdrawal of the ILS [Instrument Landing System] without replacement, thus reducing the aids available;
7. Inaccurate weather reporting, with particular reference to cloud height and visibility; and

8. The controlling authorities have not kept pace with aviation developments. An example of this is the reluctance on the part of the controllers to allow instrument let-down procedures, due to traffic congestion and the use by controllers of persuasive methods to coerce pilots to accept VFR climbs and descents.[4]

When I.C.A.O. last made an airport survey[5] it listed eight airports in Spain, all used by international airlines, which were below standard in runways and lighting. The developing tourist center of Ibiza had neither radar nor ILS, though numerous flights had to use the airport at night or in foggy conditions.

The I.C.A.O. found six international airports in France with inadequate equipment and landing aids, including such tourist airports as Cherbourg, Deauville and Lyons. In Italy, according to the survey, there were shortcomings at Naples, Venice, Milan and Bari. Yet these are all relatively busy places. If the airport authorities concerned will not install the equipment voluntarily, then the airlines which use these airports must exert pressure to see that they do, under threat of withdrawal of services.

To list the needless obstructions and hazards that lie along airport approaches or are, surprisingly often, within the airport perimeter itself would show how extraordinarily casual some authorities are on this point. The assumption seems to be that the unusual never happens, that a plane never dips below its proper glide path, never lands short, never skids off a runway in wet weather, never overshoots. Absurd as it may seem, the accident records contain numerous examples of crashes caused by airport obstructions whose potential danger must have been glaringly obvious to

[4] IFALPA, 23rd Conference, Oslo, 1968.
[5] I.C.A.O. Air Navigational Plan and Regional Survey, 11th Edition, February 1967; IFALPA, 23rd Conference, Oslo, 1968.

The Overcrowded Airports

any airport official taking a serious look at his safety precautions.

A number of accidents have been caused by planes striking trees along their glide path for landing, and quite close to the perimeter of an airport, where an aircraft's *proper* altitude is little more than a few hundred feet up in any case. It needs only a slight altimeter error for this potential danger to become an actual one. One might have supposed that action would have been taken to see that timber was better grown elsewhere. It is high time that authorities took a critical look at trees, buildings and other obstructions along the near section of the glide path, and took steps to remove the dangers. Pylons, telegraph wires, even neon advertising signs, also exist in potentially dangerous places, and should never be allowed.

A good many of these obstacles are obvious from a passenger window, as a plane comes in for its touchdown. Less obvious are such man-made dangers as the ditches, mounds and dikes that lie alongside or at the end of the runways.

Then there are the airport's own installations. One can often see, for instance, the small brick or concrete building which houses the main ILS transmitter equipment situated quite near the runways. It needs only a skid, an overshoot, a mishap like a punctured tire or a fractured undercarriage for this sort of structure, and the ditches, mounds and so on, to become menacing.

If the plane runs into these with any force, then what might have been a minor incident can become a real disaster. The plane's wing fuel tanks may fracture and in a few moments there would be enough spillage to cause a lethal blaze.

These hazards cannot simply be blamed on lack of funds. Transmitter buildings can be put underground. Airports complain of lack of car-parking space, but resolutely continue to build upward instead of downward to provide underground space. Ground near runways can be leveled at little cost.

The topic of obstructions crops up regularly at every conference of the airline pilots' associations. At a recent safety forum in Washington, organized by the American pilots' association, an example was given of the strange and dangerous incidents than can happen when vigilance slackens. An airport director—Foster V. Jones, of Louisville, Kentucky—told how he had found a contractor building a retaining wall in a ten-foot drainage ditch just off the end of the runway. Astonishingly enough, it turned out to have been on the orders of F.A.A. "I know if we had let that construction stand and someone went off the end with a 707—and this can happen—there would not have been much left but a ball," he said.[6]

At an earlier forum, Homer Mouden, chairman of the pilots' Airworthiness and Performance Committee, had described the hazards of landing at Kansas City, whose airport had a very bad reputation among pilots. After an overshoot accident involving a big jet, the speaker said, the board of inquiry learned just what sort of aerobatics a pilot had to perform to get in there. He had to land on the seven-thousand-foot runway (which is very short for a four-engine jet) with a downward slope of 0.8 percent (which makes pulling up that much harder).

All this has to be accomplished after the pilot has had to bring the airplane over a hill and across the approach side. Then, at the far end of the runway he has an artificial hill which the CAA put in, and a road and a dike to stop him in case everything hasn't worked. You all see various types of "Kansas Cities" all over the country.[7]

Yet the landing of jets at this grim travesty of an airport had again been approved—like the ditch at Louisville—by

[6] A.L.P.A. Air Safety Forum, 1967.
[7] A.L.P.A. Air Safety Forum, 1965.

the Federal Aviation Authority, one of the bodies which are supposed to set safety standards! It causes wider concern to reflect that the F.A.A., judged against the standards set by national aviation authorities in other countries, is regarded as a relatively conscientious body. And similar obstacles, jeopardizing flight safety and passengers' lives, can be seen in Britain, France, Italy, Holland, India, Indonesia, Egypt, Spain and Germany, just to name a few countries where you'll find examples of "carefully architected negligence."

Perhaps the most chronic deficiency of the so-called modern airport is that its runways in many cases are not long enough. It is possibly the factor that causes the pilots more heart searching than any other. At the biggest of the international airports, of course, like those at London, Paris, New York and Chicago, the runways offer plenty of room in all normal circumstances. But even at these, when making a poorly calculated touchdown in wet or icy weather, pilots and passengers have been given some anxious moments.

The "overshoot"—when a plane fails to stop before the end of the runway—has plagued the industry ever since the creation of the first sizable passenger aircraft. Every year numerous overshoots still occur, leading often to damaged aircraft and, occasionally, injury and loss of life. Because of the great danger of a fuel fire—whatever the degree of impact with an obstruction—it is always a mishap that must be regarded as critical.

Overshoots are caused by various factors: worn tires, a strong, gusty tail- or cross-wind, aquaplaning or, less commonly, an undercarriage collapse, when the plane skids forward virtually out of control. On occasion the pilot is unable to get his undercarriage down because of mechanical or hydraulic fault, and then there is no alternative but a belly-landing on a foam-covered runway.[8] Incidentally, in a number of these cases, the plane concerned is diverted to an out-

[8] F.A.A. Incidents Reports, Lloyds Aircraft Accident Lists.

of-the-way airport. Although the rescue services at such an airport will be limited, its perimeter will be free of buildings and other obstructions—a tacit admission by airport authorities of the hazards that obstruct the landing zone at the big airports.

Short runways are often responsible for "undershooting." Pilots, when they realize they are approaching a strip with only marginally safe stopping distance, naturally try to give themselves more space by putting the undercarriage on the ground as near to the threshold of the runway as possible. To misjudge the landing and touch down before the concrete (perhaps on soft earth) may result in a severe shaking, damage or even disaster. Most runways have what is known as a "lip"—a difference in height between the runway surface and the prerunway earth. Many a plane that landed short has had its undercarriage ripped off by this lip.

Such accidents have appeared frequently in the accident lists even though pilots have to stick as far as possible to a regulation glide path, designed to bring them in to an actual touchdown somewhere between 1,000 and 1,500 feet down the runway. For the passenger, the speed of the aircraft at this point (around 130 miles per hour in a big jet) is deceptive. He may have the impression that the touchdown has been made only a few yards beyond the broad yellow strips that mark the threshold. In fact, around 15 percent of the length of an average runway will already have swept below the wings of the aircraft before its wheels touch.

This is a sensible precaution in avoiding undershoots. But a pilot cannot have it both ways. If he has had to allow a margin at the beginning he will not have it at the end.

What can be done about runways? The time lag in planning is now so marked and the pace of advance so swift that we will seldom see the sort of runway that meets ideal safety criteria. It is idle to demand that every sizable airport should have runways that the reasonably skillful pilot can use comfortably and without strain.

A great deal can be done without massive expenditure. The American Air Line Pilots Association has drawn up a scheme which would ease the problem of short runways. At each end of a typical 7,500-foot runway there would be an "overrun area" compacted hard enough to withstand the weight of the heaviest plane. Beyond this, again at each end, there would be a "clear zone," free of obstructions and flattened to runway level.

By cutting grooves in the runway so that the surface is roughened and water has a better chance to disperse, the risk of aquaplaning can be minimized. Every airport knows about these braking problems. But the fact is that Chicago's O'Hare, New York's Kennedy, Washington National, Los Angeles International and Kansas City Municipal are the only airports in the world which thus far have taken the precaution of "grooving."

The problem of arresting a plane that is going beyond the overshoot boundary into danger has not yet had enough serious research. When a plane is out of control, the means of arrest must be gentle and elastic, yet capable of holding uncontrolled weight of many tons at maybe 100 miles per hour. Nets have been suggested, and experiments have been made with gravel pits at the end of runways. Objections can be raised to both. Obviously a device that interferes with a plane that is not in trouble is not what is required; nor one which allows an airport to forget to raise the money for a decent runway. A gravel pit, in which the loose stones or shale act as a brake on the undercarriage or the automatic electronically operated cable system, might be useful at those airports where, without it, a runaway plane would go into a lake or the sea. But these devices are largely makeshifts. The fact that they are needed at all arises largely from the false economy and backwardness in planning of the past. Though the Jumbo jets have been on the drawing boards for four years, and in service for about one, what plans have been made for them? An overshoot by a Jumbo carrying four hundred passengers and weighing two hundred tons is going to be a different and more dangerous proposition. But whatever the size of the jet, action is long overdue in this area. The idea of a clear zone at runway ends should become a standard planning concept.

Airport fire services are also in need of modernization. In a number of accidents fire tenders have been unable immediately to get near a plane that has overshot or undershot, either because of the absence of a clear zone or because the layout of the perimeter roads did not allow easy access.

At a recent accident at London Heathrow Airport,[9] the fire fighters suddenly discovered that a hydrant very near the burning aircraft was unusable. Valuable time was lost in relocating the hoses. Surely it is not too much to ask that periodic checks be made of such basic equipment to make sure that it will work when needed.

Some airport fire services still use water to extinguish aircraft fires. While it may be better than nothing to damp down flames actually on the fuselage, it is a very dubious mode of attack. If there is a large quantity of spilled fuel that has caught fire, then water is an ineffective extinguishing agent.

The foam extinguishing agent in use at bigger airports is very useful, but a great deal of it is needed and it is also old-fashioned considering the massive quantities of fuel carried in jets and the speed with which a fuel fire can take hold. Research is needed to find a chemical that will quickly take the combustible quality out of the fuel or even solidify it so that spillage is reduced. In the United States research is under way to develop a jellied fuel that jets could use.

[9] CAP 324, B.O.A.C. 707, April 8, 1968.

The Overcrowded Airports

The Shell company has produced a chemical additive which "earths" static electricity generated in the tanks and pipes during an aircraft's refueling process. This is such a common-sense precaution that it is surprising how few airports have adopted it.

The Federal Aviation Agency is also doing tests on a device called a "bomb sniffer," an instrument which is sensitive to explosives and could detect them in passengers' luggage. Perhaps a decade ago such research might have appeared melodramatic. But investigations of several crashes in recent years have shown how easily a bomb can be taken aboard a plane by an unstable person. Though it is a very rare type of accident, its catastrophic results, as well as the vastly increased numbers of people who are going to be at risk in one aircraft, indicate the need for such a device, especially after the success last year of a Palestinian guerrilla group in murdering forty-seven people in a Swissair CV 990.

Under most circumstances the business of bringing an aircraft in to land is routine procedure, and yet the average passenger has little knowledge of the way in which this is achieved. About twenty miles away from the airport, the pilot picks up the radio beam from the ILS (Instrument Landing System) localizer, which enables him to get on the correct glide path for his approach to the runway. Any variation off the correct line is indicated by the strength of the signal and the pilot adjusts his controls accordingly.

Let us suppose the pilot has been instructed by Air Traffic Control to land at "Runway 28 Right." As a rule he can identify it visually without any trouble, because he has landed on it often before, or because he has been in the copilot's seat on previous landings. If he is less familiar with the airport he will have studied the runway system from the charts. If there is still any uncertainty in his mind he can always request identification, and the controller will flash the two parallel rows of white lights which mark the correct runway by dimming them, then bringing them back to normal power. A row of green lights down the runway center also helps to ensure that the plane is on the proper line. In most landings in fair weather the pilot lines up the plane for the touchdown by visual methods, using (if one is installed) a VASIS—a Visual Approach Slope Indicator System. A line of red marker lights leads him up to the VASIS and the runway. Next, on either side of the touchdown zone just before the runway threshold, comes the VASIS. This is a bank of red and white lights, with reflectors. They are so arranged that if he is too low the pilot will see only the red lights, or, if too high, white lights. If he is on a correct glide-slope he will see the combined red and white lights.

Once he has completed the landing roll and the braking operation the pilot looks for a convenient taxiway—usually marked with green or ultraviolet lights—which will lead him off the runway to the airport buildings.

If there is low cloud, fog, or any degree of poor visibility, the pilot relies almost exclusively on his instruments and the ILS signal to within a few hundred feet of his touchdown. All airlines have minimum visual standards; a typical rule is that a pilot may not land—and must divert to another airport—unless he can positively identify the runway lights from a height of two hundred feet. Civil Aviation regulations enforce this strictly; and since so much delay and inconvenience to passengers can arise from a diversion, it is entirely proper that the airline should take some of the burden of the decision from the pilot's shoulders. However, irate passengers have been known to complain personally to the pilot about his failure to land at a particular airport in bad visibility.

In the last few years much research has gone into the

effort to produce an automatic landing system that would allow a safe touchdown in virtually any weather. The main developers are Elliott Automation and Smiths Electric in Britain, Bendix in the United States, and SNECMA in France. The British units have been tested on some Tridents and VC 10s and have been given trials in good weather conditions on a variety of flights. The American equipment is on some 707s, 727s, DC 8s and 9s.

The essence of these systems is that the plane is controlled by electronic equipment all the way down the glide path to touchdown and braking. Each system has one or two reserve circuits which are supposed to come instantly into service if any part of the unit breaks down, either at the airport or aboard the plane. The pilot can resume manual control at any time he considers it necessary.

All three systems depend upon the airport operating an ILS. The ground unit has a high-frequency transmitter which sends up a narrow but powerful radio beam from the far end of the runway, bouncing the signal off the runway and up into the air at an angle of about three degrees. All the systems follow this beam. They differ only in their method of handling the plane in the final moments before the touchdown. As the aircraft passes over the "middle marker"—a signal near the airport perimeter—further equipment comes into play to ensure an accurate touchdown. One system relies upon two magnetic cables running parallel along the ground, their width apart being roughly the same as that of the runway. Two receivers, one on either side of the plane's fuselage, pick up the magnetic signals from these cables and balance the reception strength of each. This in turn adjusts the controls so that the aircraft is aligned with the center of the runway.

All systems depend upon the radio altimeter, which comes into play when the aircraft has been brought down to an altitude of sixty feet. The ILS signal maintains the plane on the correct line, while its height is automatically adjusted through the radio altimeter.

At the last moment each of the systems "cheat" the equipment a little to ensure a gentle touchdown. The perfect landing requires a well-timed "flare-out," with the nose of the plane pointing slightly upward to ensure that the main undercarriage wheels touch the ground first. The equipment is adjusted so that it believes that the ground level is one foot below the actual surface. This means that the plane lands at a point when the unit assumes that it is still one foot up in the air.

To activate the braking action of the aircraft two of the systems rely upon signals from the undercarriage shock absorbers, which confirm that the weight of the plane is being carried. The third system starts the braking action when the aircraft crosses a radio beam directed across the runway. As it continues the landing roll it crosses further beams which ensure further braking in graduated stages.

Initially, some passengers may have qualms about entrusting themselves to an aircraft that, in the landing operation, is going to be under the control of a machine. But it should be remembered that all big aircraft, under normal conditions, travel on the automatic pilot for up to 90 percent of any journey. I feel sure that the automatic systems will quickly show an improvement in safety as well as convenience and that those pilots who have expressed doubts about automatic landings will also be convinced. But they are not yet anywhere near a really safe systems installation that would require major aircraft design modifications.

The systems should be installed on all civil aircraft as soon as possible initially for use in good and medium weather conditions (known as "Category 1" and "Category 2" weather), to help to get the public accustomed to the idea and familiarize the flight crews. The aircraft captain will remain in charge, and as a further check he will have a visual display on a screen in front of him showing the aircraft's position and attitude in relation to the runway. If anything goes wrong he can take over immediately.

Once this learning period is over, the use of automatic systems can be extended into bad-weather conditions ("Category 3"—almost-zero visibility downward and ahead). This will reduce the need for diversions and enable aircraft to make flights in weather that at present would keep them grounded.

But, this new freedom of the sky will accentuate an already acute problem. A time may come when an aircraft flies brilliantly from A to B through dense fog, only for its passengers to find that the same fog prevents them making the ten-mile bus or taxi journey from airport to city center. Only three airports in the world—Gatwick (in England), Brussels and Tokyo—have trains connecting them with the city. It throws an ironic light on the very idea of supersonic flight: Soon it will take three and a half hours to cross the Atlantic, but if road traffic goes on growing at the present rate, passengers will then spend two hours or more making the journey into central New York or London. At Los Angeles, where the present volume of road traffic gives a foretaste of the future, dozens of passengers a week miss their flights because of traffic jams.

There are, however, some encouraging signs that authorities are belatedly recognizing the need for planning. A number of American airports—like Boston, Kansas City, Dallas and Tampa—have big development plans in hand. By 1972 the Port of New York Authority will have spent a total of $375 million on improving Kennedy, La Guardia and Newark airports. This expenditure will bring these airports up to their *maximum* practical capacity to handle aircraft. Los Angeles airport is expected to be saturated by 1975, largely because of the volume of road traffic.

But there is no need to pick isolated examples. The expert estimates mentioned earlier in this chapter, which suggest that places like Zurich and Melbourne are going to become as busy as the world's currently busiest airport (O'Hare, Chicago) during the next decade, show that almost every corner of civil aviation is going to have to go through a revolution. There will be no room for half measures.

4 · The Passenger's View

Airlines go to considerable lengths to reassure passengers that safety has absolute priority over all other considerations. But it cannot be said that the airlines go out of their way to make available to the general public the information that would substantiate their concern. I have examined over one thousand annual reports of airline companies and only one of them, B.O.A.C. for the financial year 1965–66, even mentioned the crashes and other mishaps. All the rest were mute on the subject.

Because some of the bigger companies are largely stateowned there are no ordinary stockholders' meetings at which a member of the public could ask a few searching questions. Was a particular accident avoidable? If so, why did it occur? Will the chairman admit that it was due to poor engine maintenance or to the company's refusal to buy the best possible navigational aids for the crews? Even when an airline company is largely owned by private stockholders these important questions are not asked. Apparently, airline propaganda has been so effective that it is assumed nothing can be done to make air travel safer and air disasters must be accepted as acts of fate.

This is not true. It is, in fact, a tragic misconception. A thorough study of the records and the expert evidence given at crash inquiries show that most accidents can probably be attributed, directly and indirectly, to the airline companies' inadequate concern for safety. It is demonstrable, for instance, that some airlines have a *consistently* better safety record than most others, even though they are flying much the same aircraft over similar routes. The official statistics never compare one airline's records with another's so this fact has not come to light. The table of airline safety given in Chapter 5 is, so far as I know, the first such compilation.

The fact that a few airlines can maintain a nearly perfect record of safety should in itself be enough to dispel any complacency on the subject. It demolishes the idea that it really has much to do with "fate," "bad luck," or "normal losses." In most cases the operating company is responsible for the upkeep of its fleet. They usually order from the manufacturers the type of aircraft suited to their particular needs. The cost of the basic unit of fuselage and engines usually influences their attitude toward the purchase of the extra equipment that would provide the extra margin of safety. Of course, money is found for in-flight cinema equipment, stereo sound units and so on. One is left with the impression that the airlines are more anxious to create an illusion of security than to make the necessary investment to achieve it.

Let us look at some of the equipment that the passenger will find in the cabin—and I emphasize that I am describing typical equipment to be found in the aircraft of supposedly first-class airlines flying major routes and carrying millions of passengers a year.

First, the seat. Since the passenger is going to have to sit in it for many hours on a long flight, it should be reasonably comfortable, offer decent leg room, and have a reclining mechanism that works. These are the minimum requirements. But do not forget that in many accidents the seat is also going to be his sheet anchor. Its strength may mean the difference between being able to walk away from a crash and being incapacitated. In the accident reports you can read of a number of cases in which, on the shock of impact, the seats broke away from their floor fastenings. This means that the seat, with the passenger still strapped in, is flung forward to collide with a heap of other seats and people, against the forward bulkhead.

The methods used by manufacturers for securing seats in an aircraft are not adequate. In many cases, banks of three seats are secured by a flimsy floor-locking runner under the second and third seats and another runner along the side of the fuselage. An aircraft designer at Lockheed Aircraft Corporation told me that the first thing he does when he enters an aircraft on a commercial flight is to shake his seat to see if it is loose. He has refused to sit in large numbers of seats because they were insecure in their floor mounts.

Seat-strength regulations vary from one country to another. The American rules are the strictest in the world. But virtually every designer and safety officer I have spoken to about it in the United States agrees that the regulations are still not exacting enough.

At present the rules prescribe that the seat must be able to withstand a force of nine "g" in forward stress (that is, nine times the force of gravity, presuming that the aircraft went nose first into something) at the base of the seat leg and the remainder of the seat must withstand varying stresses between two "g" in the U.S.A. and four and a half "g" in the United Kingdom in an upward direction to six "g" horizontally, holding a 170-pound person. Oddly enough, the seat belt must be able to take far more—fifteen "g" with a person of the same weight strapped into it.

Let me quote from a report published by the Federal Aviation Agency:

It is suggested that there is needless injury and loss of life from impact with rigid structures in cabin/cockpit environments and in current seats, which can be eliminated by improved seat design incorporating delethalization, especially in seat arms, seatbacks, and rigid structures in and under the seatback.[1]

Current seat tie-down requirements are inadequate, especially in the vertical direction; the upward requirements, based only on the weight of the seat and its occupants, gives a vertical tie-down strength of something less than one thousand pounds. The F.A.A. report shows that the lifting force of the four legs of the passengers seated behind the seat in question may be expected to exceed this tie-down strength by a factor of two to five before fracturing. In dynamic seat testing, consideration should be given to strike loads imposed on the rear of the seats.

In considering leg strike loads, the practice of storing brief cases and small suit cases under the seat should be viewed with suspicion, since these objects can slide forward against the backs of the passenger's legs, increasing the force of leg impact against the seat, and possibly entrapping the feet under the seat.

So, in theory, the seat belt itself will hold the passenger firm up to an impact force of fifteen "g" in a forward plane. In fact there have been accidents where the United States Army Air Research unit has found that passengers have survived short impacts of up to thirty-five "g."[2] Equally there have been crashes with a lower "g" force in which people have been killed by the breakaway of seats, or by injury which has prevented their escape from ensuing fire and explosion. These lives could have been saved.

Most passenger seats are designed with an eye to convenience of removal. Some aircraft have their seats fixed on "pallets." These consist of an assembly of two or three rows of seats complete with a gangway and carpeting. This allows

[1] "Kinematic Behavior of the Human Body during Deceleration," June 1962.
[2] USAAVLABS, Technical Report 66–43.

them to be quickly removed—at the rate of about 170 seats in about thirty minutes—in order to convert the aircraft from passenger to cargo transportation or vice versa.

This again must mean that seating strength has been surrendered to airline convenience. If safety is to be given its due weight what is surely needed are stronger seats and legs welded to the main spars of the aircraft instead of today's sliding attachments to seat runners. This should be done by the manufacturers of the fuselage and not left, as it is now, to the seat manufacturers and the airlines.

In choosing a seat design, the question of weight is naturally a big consideration for the operating company. It has been calculated that a plane carrying one pound extra in equipment costs the operator about seventy-seven dollars in lost revenue in one year, because, of course, it ultimately reduces his load-carrying capacity. When he makes his choice of seat and its fastening, can we be sure that the operator has given due benefit of the doubt to safety, as against economy?

When, again, the value of rear-facing seats as an extra safety measure is raised with the airline companies one customary reply is that they do not have enough appeal for the traveling public. But the truth is that the question of weight comes in here too, since to fit them the seats and the floor unions would have to be strengthened.

The public has only limited experience with the rear-facing seats, so their skepticism is understandable. But it should be more widely known that they do offer a greater margin of safety, as proved conclusively in tests with dummies and in actual accidents.[3]

In an aircraft accident in Germany, on February 6, 1958,[4] twenty-three people lost their lives. Most of them were in

[3] (Example) I.C.A.O. Accident Digest LPI p. 43; and Crash Survival Design Guide USAAVLABS Technical Report 67-22.

[4] Ministry of Transport and Civil Aviation CAP 153.

forward-facing seats. Twenty-one others lived, most of them in rear-facing seats. This positioning naturally provides a partial bodyguard to the impact, as long as the seats are properly installed and strengthened.

The public antipathy toward rear-facing seats is largely illogical and could be dispelled by a little education about it —by the airlines, for instance. There is little or no sensation of forward movement as in a train, with the telegraph poles whipping by the window. Those travelers who would be affected by dizziness if they traveled in a rear-facing seat in a train would be surprised to find how comfortable it is in similar seating on a high-flying jet. The ground passes so slowly that the optical effect is minimal, as doctors confirm.

Virtually all the aircraft designers and safety officers I have met agree that the seating-strength regulations need to be updated. They differ only in their views of what should be done. Some think that the seat legs should be made of tubular material which would bend rather than snap under the shock of impact, thus absorbing some of its energy.

But seating strength is not the only thing in the passenger compartment that shows a culpable lack of concern. Here are excerpts from three reports published in America:

There is a shameful and needless loss of life and facial destruction in crash impacts with transportation vehicles. Man, in a vehicle, is surrounded by rigid tubes, angles, knobs, heavy door posts, sharp instruments, and heavy metal of small radius of curvature (to name a few) all designed to impact the face and head on very small areas.

This study has shown that if this environment were changed to a medium-weight deformable metal (without heavy structure directly behind it) with a radius of curvature of six to ten inches for energy attenuation and padded with one to two inches of slow-return material to contour to the bones of the face and distribute the impact load over the available area of the face, it would be impossible to produce facial and forehead fractures in crash impacts. The limit of

The Passenger's View

human tolerance would then be the forces necessary to produce brain lacerations without fracture.

As might be suspected, the weakest part of the face is the nose, which has a fracture point varying between thirty-five and eighty g. Impact force on a single zygomatic prominence of fifty to eighty g will produce compound fractures of the arches. Condyles of the mandible will be fractured by forces of between seventy and one hundred g applied on the tip of the chin. The teeth and maxilla can withstand forces of more than one hundred and fifty g if applied to a contoured area of about four square inches. The anterior surface of the cranium (forehead) is more rugged, requiring forces ranging between one hundred and twenty and one hundred and eighty g (applied to one square inch) to produce fracture. Utilizing three to four inches of the forehead area as the impact point raised the tolerance to as high as three hundred g in some tests. Automobile accident injuries studied here established that blows to the face in excess of thirty g produce unconsciousness (fifteen minutes to two hours) with or without fractures. Variations were noted between the fracture tolerances of different heads, but these variations did not correlate with age.

Airline-seat and aircraft manufacturers should design all structures surrounding the passengers to deform with head impacts of forty feet/sec and not exceed this thirty-g figure. In addition, satisfactory padding should be provided to distribute the impact load over as much facial area as possible.

Designers of other transportation vehicles should strive not to exceed forty g since temporary unconsciousness is not such a major concern in escape.[5]

Padding of less than one-inch thickness on a rigid structure offers little or no protection during crash impact. One inch of rigid, slow-return material similar to Koroseal H334 greatly reduces impact g forces (to three hundred g) and distributes the impact load over the contours of the face when used as protective padding on rigid structure, but is still borderline for survival without head injury. At impact velocities of fifteen and thirty feet/sec against rigid structure padded with materials even six inches thick, uncon-

[5] "Tolerance of the Human Face to Crash Impact," Office of Aviation Medicine, Federal Aviation Agency, July 1965.

sciousness, concussion, and/or fatal head injuries will be produced. Head impacts at greater velocities would increase the seriousness of the injury. Underlying structures must be redesigned to deform and dissipate the energy of head impact. A combination of deforming "metal" to dissipate energy and firm padding to distribute pressure forces over the contour of the facial bones may be used successfully in preventing head injuries or even unconsciousness.[6]

Impact tests against the eight airline seats studied show that portions of some have good deforming characteristics. The most lethal design features were found to be tubular construction (round or square), nondeforming serving trays, rigid seat arms protruding rearward between the seats, and excessive break-over forces. An analysis of this series of head impacts based on earlier work shows that thirty per cent would have been fatal, eighty per cent would have produced facial fractures, ninety-seven per cent would have rendered the passengers unconscious, and only three per cent would have produced no injuries or unconsciousness.

This study shows that the following design requirements are necessary to improve the crash-safety design of seats:

a. Tubular construction should only be used in areas where it cannot cause injury.
b. Serving trays and seat backs should be molded of light aluminium sheet or other material that will deform at loads less than thirty g and contour itself to the head and face.
c. All exposed areas should be padded with sufficient slow-return foam to aid distribution of the impact force over the contour of the face.
d. The forces necessary to break the seat back forward should be reduced.
e. The lethal characteristics of seat arms should be eliminated.[7]

[6] "Evaluation of Various Padding Materials for Crash Protection," Office of Aviation Medicine, Federal Aviation Agency, December 1966.
[7] "Evaluation of Head and Face Injury Potential of Current Airline Seats during Crash Decelerations," Office of Aviation Medicine, Federal Aviation Agency, June 1966.

The Passenger's View

Glance around the cabin on almost any flight and what do you see? Pieces of the interior fittings loose, badly fitted, missing altogether, or in the process of falling off. Then there's the back of the seat in front of you that is most unlikely to have any padding that will absorb impact. Crash researchers have found that there would have to be at least two inches of padding to ensure relative safety.[8] The seat back is likely to be very close to you, in the tourist accommodation at least, because the operators desire to get as many seats on the plane as possible; and remembering the I.A.T.A. regulations regarding maximum passenger space, there is a strong risk that you would crack your head against it in sudden turbulence or other impact. There are usually plenty of other sharp edges of fittings to be seen. On some B.E.A. planes you will find, between the front row of seats (which face backward) and the second row of seats (which face forward), an immovable table-cum-magazine-holder, which could cause serious injury to occupants of those seats should the plane dive suddenly, be involved in an accident, or have to ditch in water. It would, in fact, be impossible for these passengers to assume the bent-forward ditching posture without having their heads on, under or against this table.

There are often examples of unsecured airline and passengers' gear to be seen and, though it is hard to believe such carelessness possible, cabin crews on certain airlines still apparently allow passengers to place heavy articles on the overhead racks. It is not uncommon to find personal baggage stowed on spare seats, even on the ones that offer access to the emergency exits—which are all too narrow anyway.

These practices are obviously dangerous. Should anything untoward happen during takeoff or landing, or during a spell of turbulence, unsecured objects do fly about with some

[8] "Evaluation of Head and Face Injury Potential of Current Airline Seats," U.S. Civil Aeromedic Institute.

force. There have been a good many cases of serious injury caused by them.

But there is another reason why these malpractices and inadequate fittings are highly important. The United States Army has made a study of a large number of accidents, and their findings show that two thirds of all aircraft crashes are what I shall call "walk-away," or "survivable," accidents.[9] Only a small minority should be total disasters—head on into a mountain, say, or catastrophic breakup in flight—in which there was no hope for anyone aboard.

Most accidents do leave a very good chance of escape, provided that the passenger is not too hurt by the initial impact to take advantage of it. This means that he must have a seat that is well enough anchored not to break away on impact. It means that the seats must be designed with a scientific thoroughness and with an imaginative awareness of what happens in a crash.[10] When a British Midlands Airlines Argonaut crashed in Stockport, England, in June 1967, seventy-two people died. Thirty-five of them died from burns,[11] a fact that shows that they survived the initial impact. But a pathologist said that some of the passengers had had the lower part of their legs crushed as a result of the fracture or bending of strengthening bars in the seats following a general "seat breakaway," and this, the inquiry was told, "prevented the passengers from escaping or trying to escape."

In a good many of the "walk-away" accidents the impact and damage to the aircraft can be relatively slight. The pilot, let us say, has aborted a takeoff. He has got up speed down the runway, but then, for some reason, he has decided that

---

[9] Crash Loads Environment Study, Mechanics Research, Inc., for F.A.A.

[10] Crash Loads Environment Study, Mechanics Research Inc. for F.A.A.

[11] Board of Trade Report CAP 302.

he has not enough power to get airborne, or some last-moment fault has been noticed. The pilot will then use engine reverse thrust and his brakes to try to bring his aircraft to a safely controlled halt. But such is the lack of open, emergency space around most airports—and so often are there needless obstructions left even within the runway areas—that he may be fortunate to do so without a severe jolting or even impact damage to the aircraft.

When this has happened it is too late for the stewardess to regret having allowed a passenger to place an attaché case in the overhead rack. The jolting or the impact have already flung it down the cabin, very likely causing an injury that will reduce someone's chances of what should have been a relatively safe and simple "walk-away." Nor is it the time to recall that coats and other gear left on empty seats can litter the gangway to an emergency exit. They are already doing so, creating extra difficulties, when there is simply no margin for them. On an international flight I have seen a woman passenger allowed aboard with four fully laden shopping baskets. One was placed on the hatracks, one under the seat, and the other two were on the seat next to her. The Air Canada hostess seemed to think that this was quite a safe and normal form of stowage. When I suggested that it was not and that it was the sort of thing that could be troublesome in an emergency, she said there was nothing she could do about it.

There is no time to deal with these sorts of obstacles and hindrances even in the sort of low-impact emergency I have described. An aircraft taking off on a trans-Atlantic flight will be carrying up to twenty thousand gallons of highly volatile fuel, and the fuel lines, in some planes, do not need much stress to rupture. That aircraft must be evacuated calmly but quickly, and this is certainly quite possible given good conditions with all the emergency exits free and usable. Yet we invariably picture an evacuation taking place in near-perfect conditions, with no coats or shopping baskets to hinder our movements, and no needless personal injury to incapacitate us. The airlines' safety measures seem to be based largely on this assumption.

But what happens in practice? Let us look at a fairly typical "walk-away" accident and see some of the things that can happen. Flight 861, a DC 8 of Trans-Canada Airlines (now Air Canada), was setting off from London on a foggy November evening in 1963.[12] In the middle of his takeoff run, the pilot thought that he was getting no response from the controls. Despite braking procedures, the plane overran the end of the runway at a high speed. Though the captain maintained his braking, the aircraft ran over the raised surface of the airport perimeter road, went through a fence, broke its nose landing gear when it hit a concrete verge, struck the ILS localizer with its wing, slid across a ditch eight feet wide and five feet deep, where the main landing gear collapsed and came to rest in a cabbage field half a mile from the end of the runway. There was on board a crew of seven and ninety passengers. The plane was carrying sixteen thousand gallons of JP 4 fuel, which is less safe than kerosene in accidents like this.[13]

When the aircraft came to a rest small fires were burning in the Number 1 and Number 2 engines. Due to crash damage the cocks which shut off the fuel would not work to their full extent. Attempts were made to put out the fires with portable extinguishers, but two out of three of these did not work. Though alerted within a couple of minutes of the crash, the airport fire engines did not reach the plane until twenty-three minutes after it, because of difficulty of finding a negotiable route to the field—though this was only eight hundred yards from the end of the airport runway.

[12] I.C.A.O. Accident Digest No. 11, p. 47.
[13] Report of the Working Party on Aviation Kerosene and Wide-cut Gasoline, Ministry of Aviation, CAP 177.

The Passenger's View

The official report notes:

> The emergency lighting was not bright enough to enable passengers to read instructions for opening exits or, in some cases, to release their safety belts quickly. This caused some delay, but the passengers were evacuated in an orderly manner. . . . An unnecessary waste of time resulted for some passengers from an attempt to push out an emergency exit designed to be pulled in. . . . The reaction of the passengers to the need to evacuate the aircraft quickly varied from hysterical anxiety at the one extreme to concern only for their hand baggage at the other.

There were no fatalities in this accident, and only one crew member and four passengers were injured. Since the engine fires had burned for twenty minutes, the inquiry seemed to think that those aboard had been very lucky. "No injury was sustained as a result of a shortage of emergency exits, but if the fires had developed rapidly it might well have been otherwise," says the report.

The provision of emergency exits is obviously one of the most important aspects of air safety, yet it is one of the most neglected.[14] Nearly every type of aircraft has a different method of operating these exits. Some must be pushed outward, some pulled inward. With some, a handle must be turned in the correct direction; with others you have to pull out a rope that removes a retaining bolt. With still others you must push, then pull the exit hatch inside the cabin "and place it on the seat next to the exit"—assuming that the seat is not occupied, that is, and that there will be some way of placing it so that it does not block the gangway; and assuming that the exit door works. I was with four other people at an aircraft factory in England late in 1969, inspecting a Trident I that was being overhauled. I requested permission to see how the forward starboard emergency exit worked. It

[14] Air Line Pilots Association Air Safety Forums 1966 and 1967.

was of the type which required that you turn the handle, pull it toward you and raise the door on its runners over your head. The only difficulty was that if you pulled it toward you at a slight angle it jammed; when I did this it took three technicians with screwdrivers five minutes to make it operable again. The plane was in for overhaul, but I did not see any indication that work was to be done on the door.

What disturbs me about these things is the feeling I get that no plane manufacturer or airline board has ever sat down and considered what are the essentials if you want to ensure speedy evacuation of an aircraft in the confused condition of an emergency.

What is so often lacking is not so much sophisticated equipment as the most elementary guidance on how to get out of a crashed aircraft quickly and safely. I.C.A.O. regulations state that passengers must be shown before takeoff how to operate emergency exits as well as seat belts and life jackets.[15] I have yet to travel on a passenger aircraft where this is done, so I presume that the rule is tacitly broken by all airlines.

The instruction leaflets issued are quite inadequate as a way of informing the average passenger. Most folders not only deal with the exits of the aircraft you are on but confusingly explain the evacuation methods of all other types of aircraft in their fleet, all of them different, and all of them usually on international flights, available in four languages. So the chances that a passenger will be reasonably well-informed about what to do in an emergency are slight. It is presumed, of course, that the passenger, being confronted with one of these "multi" folders, knows the type of aircraft he is on, but this is often doubtful. The British European Airways and Britannia Airways booklets also contain route

[15] Operation of Aircraft; Regulations and Recommended Practices —Annex 6 to the I.C.A.O. Convention.

maps, advertisements and other paraphernalia that have nothing to do with safety or evacuation.

The muddle of types is such that even the most air-minded travelers can be confused. One official of the Federal Aviation Agency told me of his experience when an airliner he was in crashed on landing. No one was seriously injured on impact, though the plane finished upside down. He managed to undo his seat belt, fell the few feet to the ceiling, then made his way to the emergency exit behind his seat. A queue of people quickly formed behind him but he was quite unable to find the handle that would operate the exit. Those behind him began pushing and urging him to get a move on; general panic was about to spread. After a few minutes another exit was opened and everybody got out. Being curious about his failure, my friend later returned to the aircraft, found the handle instantly and proved that it worked. He realized that when the cabin was inverted a curtain and a piece of shelving was hiding the handle from view. It obviously had not occurred to the designers or anyone else responsible for safety that planes do crash-land in awkward positions.

This was at least so obvious to one board of inquiry, investigating a crash in January 1960, that they made a special note of it.[16] A Lockheed Super Constellation of Avianca Airways, arriving at Montego Bay Airport, Jamaica, made an excessively heavy landing on the runway. It bounced into the air and landed on its back. Fires were quickly burning almost all the way around it. Of the seven crew and thirty-nine passengers, only nine persons survived. The board of inquiry made a long list of recommendations. It urged that emergency exits be provided on a scale that relates to the seat density of each compartment, and not simply to the over-all seat capacity of the aircraft; that instructions for opening emergency exits be clearly displayed and readable

[16] I.C.A.O. Accident Digest No. 12, p. 111.

whether the aircraft is in the normal position or upside down; and that compartment doors be provided with enough clearance of the door jambs to avoid jamming on impact.

The same question of exits had previously come up in 1957 at a British inquiry into the crash of a Viking aircraft of Eagle Aviation at Blackbushe Airport, England.[17] There probably had been an engine failure at takeoff, and the pilot crashed short of the runway trying to land again on one engine. Thirty-four of the thirty-five people aboard lost their lives. The inquiry quoted the one survivor as having said that "no instructions were given concerning emergency exits, their location or manipulation by any member of the crew." The inquiry said they did not intend this to be a criticism of Eagle Aviation, since there was no obligation to give such instructions (true at that time).

This was another case where the aircraft caught fire shortly after the crash. The one survivor had escaped simply because he had noticed that the window next to him was labeled "Emergency Exit—Push," and he had thus been able to open it. "It is a matter for speculation," the inquiry concluded, "whether, if instructions had been given, more of the passengers might have acted in the same way."

Another crash in the same year, 1957, again suggested that deficiencies in the evacuation system were not giving passengers a fair chance in this sort of "walk-away" accident.[18] This was when a DC 6 of Northeast Airlines, flying from New York to Miami, crashed a minute after takeoff. Of the one hundred and one people aboard the plane, twenty passengers were fatally injured, twenty-eight were seriously injured, and fifty received minor injuries.

This crash was officially held to be probably due to a number of things, including misinterpretation of instruments and loss of pilot control. It was one of those accidents which

[17] I.C.A.O. Accident Digest No. 9, p. 100.
[18] I.C.A.O. Accident Digest No. 9, p. 45.

show, paradoxically, that in some circumstances an aircraft can endure a considerable impact; and that passengers can be left with surprisingly high chances of survival—so long as someone has thought far enough ahead, and with enough imagination, to see things in a crash context. A wing tip of this DC 6 first struck small trees. Then both wings in turn struck the ground while the plane was still moving at a speed of 138 knots. The plane skidded along on its belly for about five hundred yards before coming to a halt.

Despite the force of this crash, the board concluded, there were no fatalities or serious injuries caused by the actual impact. The cabin floor remained fairly intact, before the plane caught fire, and the seats were not loosened from the floor. It was the subsequent fire, fed by three thousand gallons of fuel, which caused the many fatalities and serious injuries.

"There was obviously some deformation of the fuselage," the inquiry report suggested, "during the slide which jammed the main cabin door, and possibly other exits as well. Civil Air Regulations require that aircraft doors and emergency exits be constructed to minimize jamming in minor crashes. Obviously, in impacts and ground movement of the magnitude of this accident there is no aircraft construction that would prevent fuselage deformation and consequent jamming of exits."

The absence of lights in the cabin after the aircraft came to a stop undoubtedly hindered the evacuation of many occupants. The board was concerned about the difficulties experienced in evacuating passengers after the aircraft came to rest. The jamming of the main door and the nonoperation of the automatic lighting system apparently hindered speedy evacuation. The board is studying this problem from the standpoint of adequacy of the regulations and their application.

No doubt the board did go ahead and "study the problem"

in all good faith. But note the date of this piece of assurance —1957. What does seem extraordinary as well as tragic is the length of time and the number of needless accidents it takes to make the airline industry fully aware of even a major safety defect—or at least to get something effective done about it.

It is hard to say precisely how many people have lost their lives through inadequate emergency exits in the twelve years since that inquiry. In many cases it has been at least a contributing factor to the number of fatalities. One need only mention some of the bigger crashes where inquiry evidence suggested that the evacuation problem was a major factor in the size of the death toll.

In July 1961, an Ilyushin airliner of Ceskoslovenske Aerolinie crashed near Casablanca, Morocco, while attempting to land in poor weather.[19] All seventy-two people aboard died in the crash. The official accident report includes the statement: "When the police arrived at the scene twenty-five minutes after the crash . . . calls for help were heard coming from the wreckage, and an attempt was made to rescue the passengers, but a fire started, and it was impossible to continue operations."

In November 1961, a Lockheed Constellation of Imperial Airlines crashed at Richmond, Virginia, with the loss of seventy-seven lives.[20] It was reported that all the victims had died through carbon-monoxide poisoning. Only the captain and the flight engineer escaped from the burning wreckage. "Subsequent investigation," says part of the inquiry report, "showed that many of the passengers had left their seats after the impact and had attempted to evacuate the aircraft. . . . The largest group of passengers was found near the main-cabin entrance door, which either had been jammed by the ground impact or by trees and debris, which were piled

---

[19] I.C.A.O. Accident Digest No. 13, p. 129.
[20] I.C.A.O. Accident Digest No. 13, p. 251.

up against the fuselage. There was no evidence that attempts had been made to use any of the emergency over-the-wing window exits. No positive evidence of impact injuries to the passengers was found."

Again we see the same pattern. A large number of passengers managed to survive what was quite a severe crash (the plane hit two groups of trees when about fifty feet up and moving at a speed of 90 knots) without any disabling injury. Yet it did them no good. Why? Again there was the familiar jamming of at least one exit. And why did at least some of them not try the over-the-wing exits? Had these been pointed out to the passengers before takeoff?

The sorry story of these crashes continued and, eventually, *Time* magazine took notice of it. "The trouble with today's passenger-crammed jets," it wrote on September 29, 1967, "is that too many people do not get a chance to walk away—even from crashes that the Federal Aviation Administration classifies as 'survivable.'" It recalled two other major accidents, besides those I have mentioned. There was the crash of a United Air Lines DC 8 which swerved off the runway at Denver, Colorado, in July 1961 and hit a concrete obstruction (note, by the way, the presence of this hazard near a runway). Sixteen people suffocated, probably because the emergency exits jammed after fuel from ruptured lines fed a fire in the cabin. Then, in November 1965, there was a similar accident to a United Air Lines Boeing 727 jet at Salt Lake City (the one we dealt with in the first chapter), in which forty-three people died, most of them needlessly. The F.A.A. itself estimated that between 1961 and 1967 more than two hundred and seventy persons had died in survivable landing or takeoff accidents.

This old and neglected problem of emergency exits *must* be tackled seriously and solved, if we are not to run into some stunning catastrophes in the age of the Jumbo jets with their 425, or even 900, passengers aboard. The airlines and the manufacturers, clearly, have not tackled it with the necessary vigor, with a real all-out determination to find an answer; otherwise we should not have this regular toll in the "survivable" crashes. If they have not solved it with the present generation of jets, why should we believe that it will be that much better in the Jumbo era?

But with the Jumbos it is worse—double exit doors, for sure, but a drop of 16 feet 3 inches to the ground[21] (if the landing gear hasn't snapped off), and the only method of escape at the moment in the Boeing 747 is double slides, slides similar to a type that has proved faulty under crash conditions in the past.

Even now there is no sign that all humbug on the subject has been put aside, even by the most safety-conscious agencies, those of the United States government. In the latter part of 1968, though it came late, the F.A.A. and the Flight Safety Directorate at the British Ministry of Aviation did lower the minimum evacuation standard to ninety seconds. The manufacturer must now demonstrate to the inspectors that it is possible to evacuate the plane in this time with a full passenger load, a full crew, and half the exits designated "unusable" before he gets the certificate.[22]

Let us admit that it is a small, though timid, step forward, and we should be grateful for that. But is it, honestly, good enough? I have seen pictures of one of those evacuation tests —one-act theater, one might call it—in progress. The sun is shining; the test plane is neatly parked on a wide stretch of well-swept concrete. A very orderly group of "passengers" stands on the wings, each patiently awaiting his turn to be helped down.

No smoke, no fumes, no flames, no injuries, no stunned minds, nothing blocking the exits or passageways being used. Of course they will always manage the ninety-second

---

[21] The Boeing Company, "747 at the Airport."
[22] Aeronautics and Space. Title No. 14. Docket 7422.

time limit; they might even cut it down substantially. But what on earth does this prove? So the desperate illusion continues. And so it will continue until passengers rebel and demand an end to smooth assurances and get, instead, a sense of honest reality in the safety regulations. Surely the lesson of all those tragic accidents I have quoted above, and of all others like them, is that the passengers could not get out at all *under crash conditions*. Ninety seconds meant nothing to those who died at Denver, Salt Lake City, Richmond, Casablanca, Stockport, London, Rome, and the many other places where these types of crashes have occurred.

The effect of this kind of drill is to reassure the manufacturers—and the airlines buying their products—that they will not have to make serious modifications to their fuselage designs. Yet this is exactly what is needed to meet crash conditions, and aircraft designers have assured me *it can be done;* but it must be done with an initial concept of a new design in order to cost the least money.

It is quite obvious that the fuselage of a crashed aircraft is going to experience distortion by stress and heat that is likely to cause the jamming of exits. But it is hard to believe that technologists who have solved the highly advanced problems of heat-and-stress metallurgy needed to explore outer space cannot devise emergency hatches and doors which can resist distortion, or which open automatically after a crash, or which are exploded out during a crash (with some failsafe device to ensure that it does not happen accidentally). To be realistic, all exit doors should and could be exploded open before the twisting fuselage jams them tight and before the plane comes to a halt.[23] Only then will the nervous passenger realize that all he has to do is to get through those open holes to survive.

Experience also shows that exits do get jammed by trees, rocks, and other obstructions outside; or because the fuse-

[23] Air Line Pilots Association Air Safety Forum 1966.

lage is lying on its side. This is another reason why the manufacturers should be obliged to provide more doors. The regulations should insist on them the whole length of the fuselage at distances of no more than *five feet apart*.

Passengers must be shown how the exits work before take-off as in the I.C.A.O. regulations:

4.2.7.5. An operator shall ensure that all crew members are instructed and periodically examined in the use of the emergency and life-saving equipment required to be carried, and that they are drilled in emergency evacuation of the aircraft used.

4.2.8. An operator shall ensure that passengers are made familiar with the location and use of:

   (a) safety belts;
   (b) emergency exits;
   (c) life jackets, if the carriage of life jackets is prescribed;
   (d) oxygen dispensing equipment, if the provision of oxygen for the use of passengers is prescribed; and
   (e) other emergency equipment provided for individual use.

4.2.8.1. An operator shall inform the passengers of the location and general manner of use of the principal emergency equipment carried for collective use.

4.2.8.2. In an emergency during flight passengers shall be instructed in such emergency action as may be appropriate to the circumstances.

Apparently some airline people think that this would cause needless anxiety to nervous passengers; or else they casually suggest that "the steward would handle it in an emergency." Both reasons seem nonsense to me. Demonstration of exit routine—ten times more effective than simply getting people to read one of those obscure leaflets on the subject—is surely no more alarming than the drill for oxygen masks and life jackets to which we are already accustomed on long flights. Eventually it comes to be accepted as

The Passenger's View

sensible and normal. Of course, it is absurd to suggest that one of the crew is going to be on hand to do the opening in an emergency. I cannot speak for these "nervous passengers"; but I find it more frightening to think that, after a crash, the passenger who happens to be at the front of the queue at the emergency exit is going to fumble with the door mechanism.

The safety officers, designers, and aerodynamicists I have spoken to in a number of countries nearly all agree that the number of emergency exits on civil airliners today is inadequate. Only a few disagreed, and their argument is usually this: The exits would be adequate if only there were a really safe fuel. A good case can be made for this view. It is the fire and subsequent smoke and fumes hazards that are the omnipresent passenger risk in these "survivable" accidents; and I shall deal in Chapter Five with how some airlines invite an extra measure of risk by using low-flashpoint fuel. But which is to come first—the safe fuel or the adequate exits? Both have been argued about for years, and still there is no international regulation that deals effectively with either hazard.

The new F.A.A. regulations put exits into various categories: some are at floor level, some are wide, others are narrow, but basically the lack of concern is still there. A Boeing 707, capable of carrying 178 people, still has to have only eight emergency exits (the rules allow entrance doors to be classed as emergency exits). That means only four to each side of the aircraft—two of them at floor level and measuring not less than 24 inches wide by 48 inches high; and the other two located over the wing and measuring 20 by 36 inches.

A well-known and widely used aircraft like the Douglas DC 8, has already been critically looked at by a board of accident inquiry which recommended that the Douglas Aircraft Company conduct a study to improve the distribution of emergency exits to allow for speedy evacuation, following an accident wherein the plane finished up in the ocean, not far from shore.[24] Yet the new rules do nothing to correct this situation. When a DC 8 of the South American airline Panair do Brasil crashed, on August 20, 1962, the inquiry found that probably the placing of all four of the emergency exits in the central part of the fuselage "hampered the evacuation, as the number of passengers (ninety-four) was considerable." The plane ran off the runway, across a road surrounding the airport, over a retaining wall into the sea and finished up about one hundred yards from the shore. The water prevented the use of all exits except those over the wings. Now, if the exits on the DC 8 are badly sited, then one could also make the same criticism of those of the Boeing 707, Boeing 720, Boeing 727, DC 9, DC 8 Super series, Convair 880 and 990, Vickers Super and Standard VC 10, Trident, the much criticized BAC 111, the Caravelle, the Comet 4 and the Fokker F 28. And that list covers the bulk of passenger aircraft flying today.

Nor do the new rules say anything about airlines that fill the forward part of the plane with cargo, thus making the forward escape hatches inaccessible to the passengers. This has caused difficulty in at least one accident.[25]

One good point about the rules is that the F.A.A.—and this still only applies to aircraft registered in the United States—does seem to accept as inevitable that the Jumbo jets and supersonics are going to need easier escape routes. What is called a "Type A" exit must measure not less than 72 inches by 42 inches. It must be at floor level. Unless there are two main aisles fore and aft, the exit must be placed so that there is passenger flow to it from both front and rear of the plane. There must be an unobstructed passageway at least 36 inches wide leading from each exit to the nearest

---

[24] I.C.A.O. Accident Digest No. 14, Vol. 11, p. 113.
[25] I.C.A.O. Accident Digest No. 15, Vol. 11, p. 44.

main aisle. Also required is the provision, next to each exit, of a seat which could be occupied by a steward. Finally, all exits, emergency or otherwise, must be provided with a slide (commonly made of fireproof canvas or nylon) capable of taking two parallel lines of evacuees simultaneously from Class-A exits.

The chances of any airline fitting the safe A-type exits to their existing aircraft seem remote. So on the vast majority of flights we shall still not get these sensible and basic precautions that we should have had years ago.

But why should the airlines fit Class-A exits to today's planes? The exit width itself is fine, the secondary and perhaps most important means of escape is not.

On June 26, 1970, a Trans World Airlines Boeing 747 landed under emergency conditions with an engine on fire at Kennedy Airport. Eighteen people were injured in the disastrous comedy that occurred when they tried to evacuate this plane. The 747's emergency chutes did not operate properly and people fell the 16 to 17 feet to the ground. The pilot radioed to Kennedy tower to have the normal passenger stairs brought over to the plane, stopped well away from other traffic and the terminal. The people who evacuated by the stairway were uninjured.

5 · Profits versus Safety

The present accident rate, showing only one passenger killed for approximately every two hundred million passenger miles flown by all airlines,[1] seems to make nonsense of fears concerning safety, even if one remembers that a good deal of this enormous total mileage has been notched up by jets flying the transoceanic routes, where, most of the time, there is not much occasion for things to go wrong.

But if one looks more closely at this situation, one revealing fact emerges: people generally have only a partial degree of confidence in airlines. To some extent this is understandable. Not many people can step aboard a machine that is going to fly five or six miles above the earth in the same casual manner with which they would board a train. This entirely relaxed attitude will almost certainly never come, at least not for many years, no matter how much the airlines spend on their publicity. It is a fact that 70 percent of the people living in the United States, the richest and most aviation-conscious country in the world, have never flown in their lives. It is arguable that a good number of those who do travel by air do so reluctantly and only because of time pressure, and would choose some other means of transport if they could. As the president of one of America's leading airlines once put it, in a wise and revealing remark, "Fear, not fare, restricts the airborne market." In other words, larger numbers of people do not fly, not because of the customary assumption that they cannot afford the fare, but because they do not have enough confidence in the airlines' ability to get them there safely. It seems to me a maxim

[1] I.C.A.O. figures for scheduled services in 1967.

Profits versus Safety

that could profitably be displayed at every airline headquarters.

Is the fear of flying entirely superstition or simply a matter of the "alien element"? Such irrational things must, of course, be part of it; but they are certainly not the whole story. The average newspaper reader has only the most fragmentary knowledge of airline operations, but he must wonder why he is so continually fed the most glossy image of the business.

Yet even the uninformed reader cannot help noticing how often, when a crash does occur, the element of "bad luck" did not enter into it. Was it pure chance that a mechanic left an important bolt out of the flying controls and that no one was around to notice it? No, in some companies this does not seem to happen at all. Was it chance that the passengers on this or that crashed plane did not know how to work the emergency exits? No.

Despite all the technicalities and the soothing statistics, the average newspaper-reading passenger should become aware that a lot of these so-called accidents need not have happened if only someone had made a serious enough effort to avoid them. There are positive—and avoidable—reasons why the accident rate of charter airlines is much higher than that of scheduled services. Just as significant is the fact—as the Airline Safety Table on pages 139–42 demonstrates—that a relatively small number of airlines have been *consistently* safer over a period of several years. Their success, therefore, has nothing to do with magic and—considering their variety in size and field of operation—not a great deal to do with the amount of capital available. If some can do it, then all the excuses usually made for a poor safety rate fall to the ground.

Anyone who has studied the volumes of reports of accident investigations and compared the records of the different airlines is likely to take hold of one simple lesson: *If the top management of an airline is concerned enough about safety and really commits itself to the fulfillment of its concern, then the result will stand out unmistakably in its safety record.*

Just how good or bad an airline's standards are can be tested on a number of points. Some of the most common shortcomings occur in the training and supervision of cabin crews. An airline is required by law to see that its air stewards and hostesses are instructed in the emergency drills for a crash-landing, for evacuating the aircraft, and for ditching in the sea. But the experiences of crew members themselves, as recounted at the annual flight safety forums in the United States, show that there are wide variations in the way this teaching is done.

For example, I have been told by some of British European Airways cabin crews that when you apply for a job with them you have to fill out the familiar questionnaire. One of the questions is "Can you swim?" If you put down a Yes to this question, all is well—no trials, no demonstrations are needed—until the ditching. It is *then* that the corporation and the ditched passengers learn whether you answered the question truthfully or not.

Some airlines evidently take pains to make the training serious and realistic, and to see that the cabin crew can act quickly and confidently in an emergency. But too many airlines seem to regard these drills as a formality to be got through with as little trouble as possible. The results of this policy may not become apparent until an aircraft is really in difficulty. Under training-school conditions, the hostess may well know how to open an emergency exit and lower an escape chute. But under pressure, with a large number of bewildered passengers to handle, it becomes an entirely different matter.

The result of this easygoing attitude by some airline managements is that their crews do not become truly safety-conscious, and they constantly allow regulations to be ignored. Any observant passenger will see examples of these breaches of the rules, even when he is traveling with some of the big and famous airlines.

One of the commonest examples of laxity is in the stowing of passengers' hand luggage. Some airlines are noticeably halfhearted in enforcing the rules on this important point. Their cabin crews allow heavy articles in the overhead racks and personal baggage to be left on the paths to the emergency exits.

The airline that regards these matters too casually should be reminded of the tragic example of the TWA Boeing 707 which ran disastrously into trouble at Rome airport in 1964 when her captain was trying to abort a takeoff.[2] The plane was brought safely to a halt, with no injuries to passengers and only minor damage to one engine, after it had hit a construction company's steamroller at the end of the runway. Yet within twenty seconds the aircraft was demolished by violent explosions, killing many people inside and some outside. All together, fifty-one people died. The Italian investigation board found, among other things, that seats partly blocked the wing exits, and probably the escape chutes could not be got out rapidly enough.

When emergencies can arise with such speed, it is clearly foolhardy of cabin crews to allow the slightest encumbrance near an emergency exit. But they should not bear the whole responsibility of seeing to this. It is up to the airlines—as the American Air Line Pilots Association has been insisting for some time—to require that all carry-on baggage should be put in containers or restrained in some other way, and that articles which cannot be accommodated in this fashion should not be allowed in the cabin. If the airlines can't cope with this need, then regulations must be brought in—*and enforced.*

[2] I.C.A.O. Accident Digest No. 16, Vol. III.

One reason why many airlines and their crews do not take this question of cabin safety seriously enough is again the misguided philosophy, "Don't alarm the passenger." Accidents in which people have died or been injured because they were unfamiliar with exits and other safety equipment suggest that this policy is completely wrongheaded when it comes to a real emergency—which is exactly when it does matter—and it has apparently not done a great deal to encourage people to fly with total confidence.

Rather than have their crews brief the passengers beforehand on emergency procedure, most airlines provide only an inadequate booklet. The regulations require that one cabin-crew member be carried for every fifty passengers, and some of the biggest airlines appear to take no voluntary steps to improve this ratio. It obviously means that cabin service will be slow and, much more serious, that in an emergency the amount of help and guidance available to each passenger will be strictly limited.

The efficiency of the cabin staff is clearly of such importance that they should be professionally examined and licensed by the government concerned. A high standard of competence in all matters to do with safety should be required, and licenses should be suspended for breaches of regulations, just as is the present case with flight crews. This has often been urged by pilots' associations and air-safety organizations, who know the importance of a really proficient cabin crew.

Another vital area where a management's true safety standard reveals itself is that of aircraft maintenance. Faulty maintenance causes delays to aircraft more commonly than do accidents, and one recent spot check made at an international airport showed how surprisingly frequent these defects are. The fail-safe mechanisms on aircraft are a check against a defect turning into a disaster—but not always.

119 Profits versus Safety

Below I give some examples of accidents which were at least partly attributable to technical faults, chosen at random from the official inquiry reports. It would obviously be unfair to draw any general conclusions about any one airline simply because it is mentioned here. The main purpose is to show the importance of attention to maintenance and to illustrate the kind of thing that is allowed to go wrong.

1. *United Air Lines—United States Domestic—January 19, 1955 (I.C.A.O. Accident Digest No. 7, p. 58)*

The company's Convair 340 was on a scheduled flight from Newark, New Jersey, to Lincoln, Nebraska, with some intermediate stops. It had just taken off from Des Moines—with thirty-nine passengers and crew aboard—when the pilot noticed severe buffeting. A little later he found it almost impossible to control the flight. The plane was then on what seemed to him a "stall course"—it was continuing to climb at a dangerous rate. Eventually he brought the nose down and made an emergency landing in a snow-covered field. In the landing eighteen passengers received minor injuries.

The Civil Aeronautics Board inquiry revealed, among other things, that there had been excessive play in the servo-tab controlling the elevator. This had been noticed by the maintenance crew but not properly repaired, largely because of misunderstandings in passing the job from one shift to another. The C.A.B. said in its summing-up:

The probable cause of this accident was a series of omissions made by maintenance personnel during a scheduled inspection which resulted in the release of the aircraft in an unairworthy condition and an almost complete loss of elevator control during flight.

As a result of the accident, the company increased the number of supervisory and mechanical personnel.

2. *Western Airlines—United States Domestic—February 13, 1958 (I.C.A.O. Accident Digest No. 10, p. 75)*

A Convair 240 operated by the company and carrying twenty-one people had to make a forced landing in the desert shortly after leaving Palm Springs, California. The aircraft caught fire and, though there were no fatalities, most of those aboard received injuries.

The C.A.B. investigation found that the crew's difficulties in controlling the plane in flight had probably arisen because the leading-edge section of the right wing had fallen off. During the inquiry the airline's chief mechanic appeared before the board and said that he had made a walk-around inspection of the plane, but added, "I shouldn't have to check everything." The board's report observed, among other things, that the airline had failed to make a sharp enough division of responsibility between the production side of the maintenance work and the inspection side. The "probable cause" of the accident included the failure to secure the wing section. "This improper installation was undetected because of inadequate inspection."

3. *Capital Airlines—United States Domestic—April 6, 1958 (I.C.A.O. Accident Digest No. 10, p. 113)*

The company's Viscount 700 D was about to land at Tri City Airport, Michigan, but on its final approach it suddenly crashed to the ground killing all forty-four passengers and three crew members.

The C.A.B. inquiry reported that among other things,

The Dowmic Switch that arms the stall-warning system when the aircraft is airborne was found to be malfunctioning after the crash. Examination of the switch and acceleration tests conducted on similar switches indicated strongly that the malfunction existed prior to the accident.

(A stall in an aircraft in flight is encountered when the forward motion of the plane decreases sufficiently to permit the forces of gravity to supersede. The wings lose their "lift" and the plane literally "drops out of the sky." The stall-warning device operates when the aircraft is approaching this condition.)

The Board's finding was that

the probable cause of this accident was a stall during a steep turn resulting in an over-the-top entry to a spin, at an altitude too low to effect recovery. Contributing factors were an inoperative stall-warning device, gusty winds, and possible ice accretion on the airframe.

4. *Pan American World Airways—United States International—July 12, 1959 (I.C.A.O. Accident Digest No. 11, p. 156)*

The company's Boeing 707, on a flight from New York to London, lost two of its four wheels on the main left landing gear during takeoff. A successful emergency landing was made in which four of the one hundred and two people aboard were injured.

The C.A.B. inquiry discovered that the landing-gear trunk beam of the plane had been damaged on a previous flight. The Board concluded that a cause of this accident was probably a damaged bolt, which had been part of this assembly, and either was replaced without appropriate notation in the records or was the bolt in place at the time of the accident. "Either action indicates improper maintenance practices."

As I have said, it would be unfair to draw general conclusions about any one of the airlines named above purely on the basis of these examples. All but a small number of the world's airlines appear in the accident reports in a ten-year period, through one mechanical shortcoming or another.

But with these crashes caused by maintenance defects, one lesson stands out. Safety does depend on quite junior employees of the company doing their jobs thoroughly and feeling that it matters. This obviously does not happen unless they sense that the management, from the president and chief executive downward, cares as strongly about safety as it does about showing a profit. The accident reports contain far too many examples of airlines—big ones as well as small—where slack supervision has been allowed to go unchecked for an alarmingly long period.

It is much the same with the efficiency of cabin crews. Again it is the all-pervasive company spirit that counts for most. That is why, when I see stewards allowing emergency exits to be cluttered with parcels, I have a strong suspicion that similar slack practices have crept into the workshops and every other department of the airline.

The success of an emergency evacuation, one of the two or three most important aspects of air safety, depends very much on one thing. Vernon A. Taylor, of the Flight Safety Foundation, put it aptly in addressing the Air Line Pilots Association Safety Forum in Chicago in 1965.

Leadership is the key factor in any successful evacuation. It is the ability of the cabin attendant to make the right decisions at a critical time that accounts for the many lives saved in survivable accidents.

The question of hard cash is, of course, intimately bound up with the airlines' view of their responsibility for passenger safety. Their managements maintain such secrecy about the whole subject—all the way from their strictly confidential lists of daily mechanical failures and in-flight "incidents" up to the complete failure to mention crashes in their annual reports—that it is impossible to estimate what degree of safety they genuinely aspire to.

Whenever the airlines are urged to put better equipment in the cockpit or the passenger cabin, they invariably assert that this is going to reduce earnings, because of the extra weight to be carried and the cost of the equipment itself. The final rejoinder is that increased safety would mean increased fares.

This seems to me one of the more dubious myths camouflaging the reality of air safety. Some critics of the airlines, quite well-informed about aviation economics, have suggested that the profit margins of the major airlines are already on the high side. Since most fares are agreed jointly by the airlines themselves, via an organization called I.A.T.A., and rigorous steps are taken to keep out cut-price interlopers, it is hard to say what is a fair market price for any particular flight.

We should not readily accept the major airlines' plea that better equipment is beyond their means or too much of a "luxury" to install. This is a conspicuously short-term view, of the sort that is too much inclined to depend on advertising and a good publicity "image." The public at large knows instinctively that it is not good enough, otherwise fewer of them would shy away from flying and the airlines would start to fill up those empty seats.

More thought must also be given to some other aspects of safety which are often disregarded on supposed grounds of economy. One of these is the survival equipment available to passengers when an aircraft "ditches" in the sea. In the last war it was a very common occurrence, and much was learned about how these emergencies are best survived. But in peacetime, because ditching is less common than land accidents, and no doubt because of cost factors, it has become one of the more neglected areas. B. W. Townshend, a British aviation consultant, who is possibly the leading international authority on this subject, has written:

Provision for survival in this emergency is, however, too often painfully inadequate and compares most unfavourably with that available in shipping disasters, although sea travel has a far better record of safety.

International regulations at present allow oversea flights of up to roughly three hundred miles without even *life jackets*—the presumption being that, given thirty minutes at cruising speed, an aircraft must be able to reach a suitable landing field if in trouble. Flights of up to nine hundred miles or more are permitted without the need to carry *life rafts*.

A booklet on the subject by Mr. Townshend[3] suggests how far short these requirements are, considering what really happens in a sea ditching. The gap between textbook theory and reality is the thing that airlines and government authorities seem chronically unwilling to cross.

For example, a good many airports on coasts and islands are right at the sea's edge, and planes getting into trouble on approach or takeoff not infrequently have had to ditch or have been forced down in the sea. Even if they are only a few hundred yards offshore, particularly in bad weather or at night, the immediate problems can be as tough as they might be in mid-Atlantic. People have died in accidents only a couple of hundred yards from the shore.

Secondly, if passengers have only life jackets and have to face immersion when a plane sinks, they could be forced to endure temperatures lower than 40 degrees Fahrenheit, even in temperate sea areas. In this situation the fit person may be able to stay alive for up to two hours; but others, depending on age and condition, much less. Even when num-

[3] B. W. Townshend, *Ditch or Crash-land?* (privately printed, 1965).

bers of aircraft and ships may be engaged, experience has shown conclusively that finding aircraft wreckage and survivors, even with a last-known estimated position, can be a long-drawn-out business. The effort of picking up individual passengers dispersed over an area can also take longer than theory suggests.

Townshend analyzes several ditching accidents to show the many aspects that should be improved. One of them at least is worth quoting in summarized form, since it vividly illustrates what a cabin crew has to be prepared for and the importance of survival equipment.

A Pan American DC 4 carrying sixty-nine crew and passengers, including six infants, was taking off from San Juan airport, Puerto Rico, on April 11, 1952, bound for New York, when an engine failed.[4] Despite the pilot's efforts he could not maintain height, and partly to avoid congested areas on the coast, he ditched in the sea about four and a half miles offshore. There were ten- to fifteen-foot waves at the time.

The plane was fully equipped with life jackets and had enough life rafts to carry everyone. The customary leaflet describing the location and use of the life jackets was available in each seat pocket. But, according to evidence at the official crash inquiry, passengers apparently were not orally informed, in the nine minutes between takeoff and ditching, where the life jackets were or how to use them. Nor were they told anything about life rafts.

The plane sank in about three minutes. The purser and steward shouted to passengers that life jackets were in the seat backs, then went forward to open the two left emergency exits. They climbed on to the wing and helped passengers out. Because of the lack of instruction on life-jacket use, "considerable confusion occurred."

[4] C.A.B. Accident Investigation Report 1–0026; I.C.A.O. Accident Digest No. 4, p. 58.

The captain ordered passengers to evacuate, but there was little response and he had to eject several forcibly. The first and second officers boarded the only life raft launched and took on board five passengers, the purser and steward. The captain and seven other passengers remained floating in the sea, until they were picked up thirty to sixty minutes later (the captain had been swept overboard by a wave and had been unable to return to the aircraft). Sharks were seen near survivors, but it is not known whether anyone was attacked. Amphibious rescue aircraft were over the scene within a few minutes.

Despite this, fifty-two passengers died. Townshend points out several things that could have been remedied. The life rafts, stowed in one compartment behind the pilot's cabin were not easily accessible, and they were hard to launch. Instruction on life jackets might have helped to save lives. Opening of the wrong doors—that is, the ones likely to let in most water—must have increased the sinking rate. And, though rescue efforts were so prompt, people were in the sea from thirty to sixty minutes.

The last word on the accident comes from Townshend (writing in 1965):

Although this occurred thirteen years ago, rescue facilities have made little significant advance in this period and, inasmuch as human capabilities are a predominant factor, are unlikely to do so. In some respects, such facilities have regressed, as use of flying boats and amphibious aircraft has greatly diminished. It is thus unlikely that such times can be much improved.

This point is worth pondering: eighteen years have now been allowed to pass without any "significant advance in rescue facilities." It is the sort of thing that can hardly fail to make the ordinary passenger doubt the airlines' sincerity about safety policy, research effort, and so on. The airlines cannot be unaware of the problem. They may plead that ditching is such a rarity that to make full provision for sea survival would be an unnecessary extravagance. So it may seem on the company's books. But not so to the passengers in an aircraft that develops engine trouble on an ocean crossing. In addition to the accidents, there are the "incidents" or near-ditchings to be examined and justified. Because these do not figure in reports available to the public, their frequency cannot be precisely established. But there is enough evidence to suggest that in every year a number of aircraft get into potentially serious trouble in over-water flights. In some cases these are so-called "near-coastal" or domestic flights, which are officially not even required to carry life jackets, though they may be flying as much as two hundred and fifty to three hundred miles offshore. How can we tell how sudden an emergency will be?

The same question arises with the somewhat longer over-water flights which are obliged to carry life *jackets* but not life *rafts*. Every year there are incidents in which the common sense of these regulations is made to look dubious. In a number of cases, preparations have had to be made for a ditching in the sea when an aircraft captain knows perfectly well that the survival equipment for his passengers is inadequate. His task in explaining, for instance, that people will be dependent on life jackets only when the aircraft has sunk is not one to be envied. Not unreasonably, considering that only a minority of any passenger complement consists of athletic young swimmers, these emergencies have caused at least mild panic. They are not, of course, the commonest kind of incident. But the fact that they happen at all has to be weighed in the balance.

Flight crews themselves are aware that these situations are not illusory. The International Federation of Air Line Pilots Associations, which has a worldwide membership of twenty-five thousand pilots, has recommended that life jackets for everyone should be carried on all civil aircraft. In addition, such cabin equipment as blankets, cushions and seat covers should be capable of flotation. Life rafts capable of carrying all on board should be carried by all aircraft flying over water at any stage of a flight.

Why do the airlines not regard these precautions as perfectly normal? They would very likely rely upon the time-worn argument that safety in the air has to be a compromise, and that the "weight penalty" involved in carrying this equipment would be out of proportion to the degree of risk. Of course they are right about the question of compromise. But why, in so many instances of this sort, does it always have to be tilted *against* the passenger and in favor of airline economics? The so-called weight penalty need not be excessive. A twenty-five-seat life raft weighing only 110 pounds is available; and Mr. Townshend has calculated that an aircraft can be fully equipped with rafts and the necessary survival aids and radio transmitters to go with them at a weight ratio of less than seven pounds per person.[5]

Catering for ditching risks becomes progressively more vital as aircraft get bigger. The task of rescuing three or four hundred people from the sea is disproportionately far greater than picking up, say, forty or fifty. Speed is all-important. If a Jumbo aircraft had to ditch, then only a full complement of life rafts, capable of being launched by a foolproof system, could give the passengers even a fair chance of survival.

If the airlines can let eighteen years pass "without significant advance" in ditching-survival techniques, can we be sure that they are ready for planes of the size of the Jumbo and the air-bus? The question is raised at every conference on air safety—without, so far, any adequate sign of concerted action. As has more than once been pointed out at these conferences—by responsible people in aviation, who

[5] *Aeroplane* journal, March 6, 1968

are not mere Jeremiahs—the industry as a whole can scarcely afford even one disaster with these huge planes, particularly if there were indications that there had been any neglect of safety precautions.

One of the biggest contributions that the airlines could make to air safety would be to bring about a substantial reduction in the fire risk. This is, no doubt, glaringly obvious to almost everyone in the industry. But there is also reason to doubt whether research has been pushed ahead as determinedly as is now seen to be necessary; and the policy of some companies with regard to the use of low-flashpoint fuel has been open to serious question. Here, again, the chances are that we shall be into the age of the giant aircraft before the operators have come near enough to mastering the risks of a previous era.

In an aircraft, fuel is a very intimate part of the element in which the passenger has to live. In a big jet, the main tank is located under the passenger cabin itself; it contains around seven thousand gallons of fuel, and there are similar quantities in the wing tanks on each side. The rear-engined jets have, in addition, fuel pipes running aft along the length of the passenger cabin. Other pipes carry the hydraulic fluid which actuates the brakes and the plane's control surfaces; this is also inflammable in most cases, though one company produces what is termed a "nonflammable" hydraulic fluid. Any impact, even a fairly slight one, which fractures the pipes or causes spillage from the tanks on to hot engines or elsewhere can become a menace.

A recent analysis by the United States Civil Aeronautics Board of a ten-year period of American air accidents involving fire, showed that more than half the deaths were due to fire that broke out after the crash. There were 1,161 passengers and crew involved. Of these, 488 died.

The degree of risk in these cases obviously depends a great deal on the volatility of the fuel in use, how quickly it will catch fire. This was the subject of a big controversy in aviation circles in the early 1960s, following two accidents which had heavy casualties from fuel fires.[6] The argument has not yet been resolved, but it is still of the highest relevance to the question of safety.

There are three kinds of fuel in common use in civil aviation. Gasoline is used only for piston-engined planes and is highly volatile. For turbine aircraft there is firstly kerosene, similar to domestic paraffin; it has low volatility and is the most widely used jet fuel. There is also the fuel known as JP4, or "wide-cut gasoline," more related to petrol. This has a much higher volatility than kerosene. A typical blend of JP4 has a flashpoint of 4 degrees below zero Fahrenheit, meaning that it is ignitable even though the liquid is well below freezing point.[7] Kerosene's flashpoint is specified at a minimum of 100 degrees Fahrenheit in supplies available in the United Kingdom, and 110 degrees Fahrenheit in the United States.[8]

This means in practice that kerosene will generally only form an explosive mixture in aircraft fuel tanks in relatively hot climates. JP4, on the other hand, can give off an explosive gas for a much longer period of time during ground handling and a good percentage of the flight. If kerosene does become ignited after spillage in an accident the flame will take hold relatively slowly. With JP4 the ignition and spread is much faster.[9] The controversy really began in 1960, when it became known that a number of major airlines, including Pan American, Trans World Airlines and

[6] "A Review of the Aviation Fuel Controversy," circular issued by the Air Safety Group U.K., December 1966.

[7] Ministry of Aviation Engine Research and Development Division, Specification No. D, Eng. R.D., 2486 (Issue 5).

[8] Ministry of Aviation Engine Research and Development Division —Specification No. D, Eng. R.D. 2494 (Issue 4).

[9] "A Report of the Working Party on Aviation Kerosene and Wide-Cut Gasoline," Ministry of Aviation, CAP 177.

Sabena, were beginning to use JP4. Air Canada, K.L.M. and Japan Air Lines also started using this fuel.[10]

The motives and truths of the matter were somewhat confused when the argument was at its height. A retrospective analysis of the controversy, published by the Air Safety Group, London,—an independent body—suggested that airlines that had switched to JP4 had done so because it was cheaper. Fuel prices vary, depending on local contracts says the Group; but they suggest that a worldwide operator could reduce his fuel bill by 8 percent by using JP4 rather than kerosene. They point out that the British Overseas Airways Corporation, a kerosene user, spent nearly thirty million dollars for its fuel in the year 1963–64. So the JP4 users would undoubtedly have made big savings on their budgets.

Among those who criticized the JP4 users—some of whom had claimed that there was no safety difference between the fuels—was the late Lord Brabazon of Tara, chairman of the British Air Registration Board, and a highly respected as well as an agreeably colorful figure in the aviation world. At one aviation conference in 1960 he challenged the advocates of JP4 to stand in a pool of that fuel and light a match; he would do the same in a pool of kerosene. The challenge was not taken up. When the Air Safety Group did its own demonstration for television, the kerosene only ignited with difficulty, while the JP4 took fire at the touch of a match and "flashed into a bonfire."

The two accidents brought the argument acutely to a head. On December 8, 1963,[11] a Pan American Boeing 707 was struck by lightning near Elkton, Maryland. The port wing became detached and the aircraft crashed with the loss of all eighty-one lives. The probable cause of the accident was given by the C.A.B. as a lightning-induced ignition of the fuel-air mixture in a fuel tank, resulting in an explosive dis-

[10] *A Review of the Aviation Fuel Controversy*, December 1966.
[11] I.C.A.O. Accident Digest No. 15, Vol. II, p. 121.

integration of the outer wing and loss of control. The tanks had contained a mixture of JP4 and kerosene.

The second disaster was that of the TWA Boeing at Rome, in November 1964,[12] which we have already mentioned. After striking a roller at the end of the runway, the aircraft disintegrated in a series of explosions, in which fifty-one persons died. According to the official inquiry, four fuel tanks exploded.

Further on, this report by the Air Safety Group pointed out that a number of airlines—Pan American and TWA among them—had continued to use JP4 after the Elkton disaster, despite the fact that the Civil Aeronautics Board had recommended to the Federal Aviation Agency a ban on its use.[13] Following this letter, the situation continued for approximately another twelve months, until TWA's Rome accident again challenged the good sense of this policy. Despite this, there still appears to have been a strong reluctance on the part of some companies to acknowledge the special dangers of JP4.

The Air Safety Group also suggests that there was some difficulty in establishing the key facts after the Rome crash. "It was categorically denied by TWA in London that there was any JP4 aboard, or ever had been, on that particular flight. (It always had been the practice of TWA's London office to deny their usage of JP4.)"[14] The review says that an American newspaper, quoting information said to have been obtained from TWA in the United States, had reported on the Boeing's refueling at Kennedy, Paris and Rome, and had indicated that there was JP4 in the tanks.[15] "This Press report highlighted an interesting phenomenon," the review

[12] I.C.A.O. Accident Digest No. 16, Vol. II.
[13] C.A.B. letter to F.A.A. December 17, 1963—Reproduced in Elkton Accident Report.
[14] *Aeroplane*, December 10, 1964.
[15] *New Hampshire News*, December 13, 1964, and *The New York Times*, January 8, 1965.

adds. "It seems that in the U.S.A., TWA had never attempted to conceal their usage of JP4, whereas in Europe denials seemed to be the general rule."

The sequel, as the Air Safety Group interpreted it, is also worth quoting.

On January 7th, 1965 (that is six weeks after the Rome accident) there came a dramatic and unexpected change. We were informed that Pan American were to suspend the use of JP4 with immediate effect . . . Even better news came two weeks later announcing that TWA also were to suspend the use of this fuel.

It is worth while examining the explanation put across by these two airlines in making their announcements. They both stated that there is no safety difference between the two fuels, but that they were suspending the use of JP4 out of deference to the public being educated to distrust JP4. They were referring, of course, to the determined campaign waged by the U.S. Airways Club, who had been steadily putting the facts before an influential section of the traveling public.

After the events of Elkton and Rome, we do not believe Pan Am and TWA could reasonably have been expected to state publicly that their reversion to kerosene was on account of safety. But at the same time we think that they dared not risk another tragedy which could be blamed on their fuel policies, so they cited a publicity campaign by an outside group alleged to be exploiting the fears of the flying public.

When *Flight* magazine inquired into the fuel policy of the major airlines it found that, up to the date of its issue of April 29, 1965, four were still using JP4. They were: Air Canada, Sabena (the Belgian airline), K.L.M. (Holland) and Japan Air Lines. Another former JP4 user, Pakistan International, had reverted to kerosene. The Air Safety Group notes that at least one other JP4 user, who admits to the practice when asked in Europe, has not admitted it when asked by the Airways Club in the U.S.A.

So far as I can discover, the four users mentioned above are still the only companies who admit to fueling with JP4. B.O.A.C. uses JP4 on one route section only when JP1 (kerosene) is unavailable, and the tanks are emptied of JP4 at the first stop.

Such policies have become hard to justify. To any impartial reader, the arguments about the special hazards of JP4 must carry a weight of conviction. The American Civil Aeronautics Board, with no lack of technical experts to advise it, has given its judgment against JP4. The Australian Department of Civil Aviation has outlawed its use by any Australian operator.[16]

The whole JP4 controversy has demonstrated how some airlines will stubbornly resist even informed and expert criticism of their safety standards when profit margins are at stake. It is true that research has been going ahead for some years on ways to make aircraft fuel safer. It is an open question whether the lack of candor of certain companies about the defects of JP4 will not have played its part in retarding these developments.

In the United States, the Federal Aviation Agency has sponsored a good deal of research into the production of a thickened, or "gelled," fuel which could be used by civil airlines.[17] Preliminary studies have shown that the addition of an emulsifying agent to jet fuel can produce a jellylike mass which has much improved fire-control characteristics. It is less likely to explode after a survivable crash, and even if it does catch fire, the spillage contains itself into a much smaller area. In one test the gelled fuel was found to be

[16] Supplement to Air Navigation Order, Part 106, DCA/ENG/1, July 1, 1962, Australian Department of Civil Aviation.
[17] "Evaluation of EF4–104 Emulsified Fuel in a Pratt & Whitney JT1D Engine" and F.A.A. Information Services Release No. 66/108.

96.7 percent less flammable than standard jet fuel. In a crash experiment, when fuel tanks were dropped from a height of forty feet, the containers filled with gasoline ruptured and spewed fuel outward and upward as much as twenty-five feet. Tanks containing gelled fuel also fractured, but only about a quart of the fuel was spilled.

In theory, this sounds extremely promising. In practice, the belated start on this aspect of air safety, as in a number of others, means that it will be at least five years before an emulsified fuel can be in widespread use in civil aviation.

Other methods of fire prevention include the fitting of crashproof tanks. But, for this to work properly, the designers must devise some effective way of protecting the pumps and fuel lines from fracture and spillage, and besides, this method of fuel containment is hopeless in some types of accident.

The best answer to the fire problem may prove to be the development of a fuel that will not ignite at all under crash circumstances. If this could be achieved it would be a far greater contribution to passenger confidence than any number of advertising campaigns. One large oil company has made a certain amount of progress in research on what is known as a *rheopectic* fuel, which would have these unignitable qualities. The process involves the addition of a chemical to ordinary aircraft fuel. The chemical is of such a nature that, while under pressure, it will not mix with or react upon the fuel. But should the pressure be lost, through a leak or crash-damage to the tanks, it will at once integrate with the fuel, causing it to thicken and lose its combustible character.

The urgent need to get this or some similar substance into production hardly needs stressing when we recall that we are only a step away from full commercial operation of the big air-bus and the Boeing 747. The prospect is once more that the airlines will be using the new planes before they can truly claim that they have the appropriate safety measures in a state of near-perfection. Naturally the problems have been extensively studied. But we will not be sure that the right answers will be thoroughly applied until the airlines demonstrate, without any hedging, that full priority is given to safety. There are many respects in which so far this has not been the case. Doubts about the state of preparedness for the Jumbo, for example, are reflected in the discussions among pilots and other well-informed aviation men at every contemporary safety conference.

Consider, for instance, the evacuation problem in the Boeing 747 which will carry up to 425 passengers, plus possibly thirty or forty infants. Under the conditions of a factory test, the makers have to show that this number can be got out within ninety seconds. Under postaccident conditions, going on past experiences related in the accident records, it could be a considerable feat if this timing can be approached. It is expected that some airlines will be operating them with the minimum of eight or nine stewardesses. For each to have to superintend the evacuation of around fifty passengers will be a task no less onerous than it would be on many jets operating today, though it is conceivable that the bigger complement of a Jumbo will impose disproportionately greater difficulties. Some airlines are expected to carry up to eighteen stewardesses on their Jumbos. This will be better, though it has to be remembered that half of them may be in the first-class sections.

The floor level of the 747, standing on its undercarriage, is about sixteen feet above ground[18] so that injury and possibly inability to move away from the crash scene without help would result from jumping. Boeing has installed an evacuation chute which will enable two people to slide down at a time. This appears to be a useful addition to the equipment. One hopes that rigorous efforts will be made to make

[18] The Boeing Company, "The 747 at the Airport."

them foolproof in practice, as well as in theory. Doubts on this score are perfectly reasonable in the light of past accidents and difficulties experienced in final tests before the 747 came into service.

In one recent accident, in which people died, one of the plane's chutes didn't operate because, it was later found, there was a hole in it. One might fairly presume, for example, that all escape chutes on present-day aircraft would be made of fireproof material, and that the trained stewardess would be able to operate them without hesitation. Neither is true, as the accident reports have revealed. Evacuations have been seriously held up by fire-damaged and inoperable chutes. Because of the unreliability of some chute mechanisms and the variety of release gear fitted in all types of aircraft, there have been cases of cabin staff not knowing how to operate a particular kind of chute or not even knowing where it was positioned. The size of a Jumbo passenger complement clearly allows no latitude for this kind of lapse.

Having reviewed some of the numerous safety factors that require the attention of airline managements, can we go a step further and ask which of them do it better than others? Many passengers, about to book a ticket, must have wondered whether they can increase their chances of a safe and trouble-free flight by choosing one airline rather than another. Curiously enough, the passenger can quickly get information that will help him to do comparative shopping of a minor sort—the availability of an in-flight film show or a vegetarian dinner, for instance, or details of the amount of leg room or comfort of his seat. But on the relative safety of an airline, not a word.

Even if airlines are silent about their accident figures in their annual reports, it is true that the various government aviation bodies compile the statistics for their own countries. But again they offer no way of comparing one airline with another, either in one country or internationally. It is a gap in knowledge which has two unfortunate results. It encourages the false philosophy among both airlines and passengers that airline safety is largely a haphazard business. It also means that those airlines which try harder or set aside a bigger part of their budget for safety training and equipment do not get their due share of credit, while those which seem more casual about safety undeservedly continue to pull in more passengers presumably on the strength of advertising.

To compile a reasonably fair airline safety table is difficult but not impossible. Obviously one cannot make a *direct* comparison between the accident figures of an airline regularly flying a tough-weather route over the Andes and one operating over easier terrain, or between a big jet operator and a company running small planes on short hops.

The following table was the result of about three years' work—accumulating figures from the airlines; making a card-index system of all air accidents from 1949 to December 31, 1969; compiling comparison tables; and looking for patterns or unusual characteristics that could point to safe or unsafe factors. From this compilation three patterns emerged: that some airlines have not been involved in injurious accidents; that some have been involved once or twice at the most; and that others were seemingly continually turning up in the accident lists.

This table, however, shows the leading airlines in the field of safety. They have been listed, where possible, in the order of numbers of passengers carried annually over the past twelve years. All the airlines in this statistical inquiry have in their fleets one or more aircraft of at least the size of a DC 3. There are in the United States numerous third-level operators with only smaller planes, but they are not taken into account.

The airlines in this list fall into two categories: Category

Profits versus Safety

A, wherein the company has been in operation for a period longer than twelve years and has not killed or injured a single passenger; and Category AA, wherein companies have been in operation for a period of less than twelve years but more than four, and again have not injured or killed a single passenger. The term *passenger* covers also people on the ground, third parties, who may have been involved when a plane crashed in their immediate vicinity.

If these airlines can show safety, why not the rest in the business?

THE WORLD'S SAFEST AIRLINES LIST

Position and Category	Name of Airline	Country of Registry and Type of Services	Year Started Services
1A	Japan Air Lines	Japan/International	1951
2A	Pacific Southwest	United States/Domestic	1949
3A	Transportes Aéreos Portugueses	Portugal/International	1946
4A	Air New Zealand	New Zealand/International	1940
5A	British West Indian	Trinidad & Tobago/International	1940
6A	Interflug	East Germany/International	1955
7A	Kuwait Airways	Kuwait/International	1953
8A	Tunis-Air	Tunisia/International	1948
9A	Aloha Airlines	United States/Domestic	1946
10A	Airlines of New South Wales	Australia/Domestic	1949
11A	Bahamas Airways	Bahamas/Domestic	1936
12A	Sudan Airways	Sudan/International	1947
13A	East West Airlines	Australia/Domestic	1947
14A	Arika Airways	Israel/Domestic	1950
15A	Cyprus Airways	Cyprus/Domestic	1947
16A	Nordair	Canada/Domestic	1957
17A	D.E.T.A.	Mozambique/Domestic	1937
18A	D.T.A.	Angola/Domestic	1940
19A	Ansett Mandated Airlines	Australia/Domestic	1937
20A	Airlines of South Australia	Australia/Domestic	1927
21A	Comair	South Africa/Domestic	1948
22A	Leeward Islands Air Transport	West Indies/Domestic	1956

140 Unsafe at Any Height

Position and Category	Name of Airline	Country of Registry and Type of Services	Year Started Services
23A	Air Ivoire	Ivory Coast/Domestic	1956
24A	Airlift International	United States/Charter	1954
25A	C.O.P.A.	Panama/Domestic	1944
26A	Gibraltar Airways	Gibraltar/International	1947
27A	Kar-Air	Finland/Domestic	1950
28A	Libyan National Airlines	Libya/Domestic	1952
29A	Matane Air Services	Canada/Domestic	1947
30A	Mongolian Airlines	Mongolia/Domestic	1956
31A	Northern Wings	Canada/Domestic	1945
32A	Overseas National	United States/Charter	1950
33A	P.L.U.N.A.	Uruguay/Domestic	1936
34A	Royal Air Cambodge	Cambodia/Domestic	1956
35A	SATCO	Peru/Domestic	1931
36A	Surinam Airways	Surinam/Domestic	1955
37A	T.A.B.A.	Argentina/Domestic	1956
38A	Toa Air Way	Japan/Domestic	1953
39A	Trans International	United States/Charter	1948
40A	Western Alaska Airlines	United States/Domestic	1953
41A	Wideroe's Flyveselskap	Norway/Domestic	1934
42AA	Aerolineas Peruanas	Peru/International	1957
43AA	Aerlinte Eireann	Ireland/International	1958
44AA	ScanAir	Denmark/Charter	1961
45AA	Windward Islands Airways	Dutch Antilles/Domestic	1962
46AA	Laker Airways	United Kingdom/Charter	1966
47AA	Kingdom of Libya	Libya/Domestic	1965
48AA	Malta Airways	Malta/International	1964
49AA	Air Mali	Mali/International	1960
50AA	Aero Transporti Italiani	Italy/Domestic	1964
51AA	San Francisco & Oakland Helicopter Service	United States/Domestic	1961
52AA	Papuan Airlines	Australia/Domestic	1961
53AA	Guyana Airways	Guyana/International	1959
54AA	Sierra Leone Airways	Sierra Leone/International	1958
55AA	Air Antilles	West Africa/Domestic	1957
56AA	SATENA	Colombia/Domestic	1962
57AA	Aeromaya	Mexico/Domestic	1966
58AA	Air Alpes	France/Domestic	1961
59AA	Air Djibouti	Somaliland/Domestic	1963
60AA	Air Jamaica	Jamaica/Domestic	1965
61AA	Air Mauritanie	Mauritania/Domestic	1962
62AA	Air Spain	Spain/Charter	1965

Profits versus Safety

Position and Category	Name of Airline	Country of Registry and Type of Services	Year Started Services
63AA	Belgian International	Belgium/Charter	1959
64AA	Botswana National	Botswana/Domestic	1965
65AA	Conair	Denmark/Charter	1965
66AA	Gambia Airways	Gambia/International	1966
67AA	Grondlandsfly	Greenland/Domestic	1965
68AA	Korean Airlines	South Korea/Domestic	1962
69AA	Libyan Aviation	Libya/International	1962
70AA	L.I.N.A.	Congo/Domestic	1961
71AA	Luftransport	West Germany/Charter	1958
72AA	Luxair	Luxembourg/International	1962
73AA	Martinair	Netherlands/Charter	1958
74AA	Merpati-Nusantari	Indonesia/Domestic	1964
75AA	Modern Air Transport	United States/Charter	1966
76AA	Namakwaland Lugdiens	South Africa/Domestic	1961
77AA	Naple Airlines	United States/Domestic	1962
78AA	North Canada Air	Canada/Domestic	1964
79AA	Northward Aviation	Canada/Domestic	1966
80AA	Rousseau Aviation	France/Domestic	1963
81AA	Saturn Airways	United States/Charter	1960
82AA	Somali Airlines	Somalia/Domestic	1964
83AA	Trans Caribbean Airways	United States/International	1958
84AA	Trans Europa	Spain/Charter	1965
85AA	Trans Meridian Flying Service	United Kingdom/Charter	1962
86AA	Vance International	United States/Charter	1962
87AA	VaranAir	Thailand/Charter	1965
88AA	Wardair	Canada/Charter	1961
89AA	Zambia Airways	Zambia/International	1963

Now, as we take this list a few places further we come to airlines that have been involved only in injuries to passengers within the last twelve years and/or deaths to passengers over twelve years ago. In this list, none of the airlines has been involved in accidents that have killed passengers and/or bystanders in the last twelve years.

Several points stand out in the table. The first is that the differences in routes among airlines do not appear to make as much difference as might be expected in their safety

142   Unsafe at Any Height

Malaysia-Singapore Airlines	Singapore/International	1947
Hawaiian Airlines	United States/Domestic	1929
Caribbean Atlantic	Caribbean/International	1939
Lloyd International	United Kingdom/Charter	1961
Air Guinée	Guinea/International	1961
L.A.P.	Paraguay/Domestic	1962
Trek Airways	South Africa/Charter	1960
Western Air Lines	United States/Domestic	1925
Qantas Airways	Australia/International	1920
Jugoslovenski Aerotransporti	Yugoslavia/International	1947
Alaska Airlines	United States/Domestic	1944
Lebanese International	Lebanon/International	1956
Quebecair	Canada/Domestic	1947
Aviateca	Guatemala/Domestic	1945
Braathens S.A.F.E.	Norway/Domestic	1946
Loftledir	Iceland/International	1944
Southern Airways	United States/Domestic	1949
Air Ceylon	Ceylon/International	1949
Iraqui Airways	Iraq/International	1945
Air Congo	Congo/International	1961

record. The companies with an entirely clean twelve-year safety record comprise virtually all types of passenger aircraft in existence and routes over all kinds of terrain and through all kinds of meteorological conditions.

The second conspicuous thing is that a high degree of consistency is shown to be possible over these twelve years by both big and small companies.

In a table of this sort the margin between one company's position and those immediately above and below is slight. It must be pointed out that the American domestic carrier, Delta Airlines, has a commendable safety record, and that but for a parked-plane accident in 1963, in which one person died, and a crew-training crash in 1967 that killed people in a motel, it would have nearly headed the list. Qantas Airways also has a very good record, having suffered only one fatal crash (in the late 1920s, in which pilot and passenger died) and a further sixteen people injured in an accident in Mauritius in 1960, though it has been Australia's overseas airline since 1921.

Profits versus Safety

What makes a safe airline? To find precise answers to this is not quite so easy as describing what makes a relatively unsafe one. It obviously depends on a spirit of concern about safety, strong enough to be sensed and followed by the employees. A company which takes it for granted inevitably runs into trouble.

As an example of the successful ones we can look at Japan Air Lines, an international carrier which, according to my figures, has been the world's safest airline for the past seven years, Swissair having dropped from number one position in 1963, when eighty were killed in a Caravelle accident. (Unfortunately, Swissair since then has added a further forty-seven to its list as the result of a bomb on one of its planes, the murderous handiwork of a Palestinian Arab group, and on August 21, 1970, nine people were injured in turbulence.) JAL has both a high figure of revenue-passenger miles and a record completely free of injury or fatality to its passengers. (Since confusion has occasionally arisen, it should be pointed out that this company has no connection with All-Nippon Airways, another Japanese company, one of whose Boeing 727s crashed in Tokyo Bay in February 1966.)

JAL's record deserves praise. But it is still not easy to define the quality that it possesses and that is missing from some of the companies in the lower half of the safety list. The fact that it is one of the few users of the suspect fuel JP4 persuades me to keep some reservations about it and to mark the company short of total perfection. On this score, my only conjecture can be that JAL's safety standards in other respects are so good that it has managed to avoid those accident situations where the more volatile fuel becomes a menace.

Conversations I have had with JAL's executives, pilots and several outsiders who know its operations, suggest that the management's interest in the factors that make for safety does have much to do with its good record. The company's president, Shizuma Matsuo, an aeronautical engineer by training, in 1967 ordered the adoption of a "zero-defect" program which aims to remove as many snags as possible from daily operations, ranging from mix-ups in seat bookings to delays in delivery of spare parts. This kind of attitude does appear to get employees at all levels involved in the maintenance of standards. A safety officer in one of the big American aircraft manufacturing companies (Lockheed) described the JAL engineering workshops in Tokyo as "immaculate." He had been surprised, he said, at the neatness and cleanliness of the maintenance hangars, the impression that there was not a tool in the wrong place, and pointed out that this could indicate conscientious standards on the engine work itself.

When I asked one of JAL's senior pilots if he could volunteer any reasons for the good safety record, he suggested that it was partly because the company took pains with crew training. It was concerned that rules were not only made but also observed. It also reposed a good deal of trust in the judgments of its pilots and did not, like some airlines he knew, attempt to run an individual flight from behind a desk at headquarters. The decision as to whether a flight could be safely started and concluded was left to the pilot himself; and if—as occasionally happened—this judgment of a bad-weather situation led to delay and reproaches from a few passengers, then the company always backed up the pilot, so long as his decision was seen to have been a responsible one.

On September 28, 1968, Pakistan International Airlines pilots went on strike following the dismissal of one of their captains by the company. The captain and a ground engineer were dismissed following an accident at Canton Airport, where a Boeing 707 of Pakistan International Airlines was delayed for three days.

The airline pilots met and gave two reasons for the strike. They alleged that the airline management was "forcing pilots to fly unserviceable aircraft, thus endangering the safety of aircraft passengers and crews." They went on to say that the pilot, Captain M. Saleghee, had been dismissed "when he refused to operate an unserviceable aircraft."[19]

Subsequently the dispute was resolved and the captain reinstated, but the point of the matter is, Should this situation have arisen in the first place?

I do not for a moment suggest that JAL is the only company which runs on such high standards. But they offer some tentative reasons for the company's superior placing in the safety table. It is also further evidence that if an airline's management has the right attitude regarding safety, then that insidious combination of small lapses and errors which goes to create the majority of accidents has far less chance of developing.

Finally, a word needs to be said about financial compensation by the airlines. If a passenger is injured in an accident, perhaps so seriously that he can no longer pursue his livelihood, what chance does he have of redress? Or, if he is killed, can his relatives be assured of compensation?

Regrettably, cases in which niggardly compensation has been paid have had less publicity than they deserved. One possible reason is that claims take so long to settle that they inevitably lose what news value they originally had. An airline and its insurers customarily withhold payment until the board of inquiry has announced its conclusions on an accident, and this can take up to a year or more. Interim payments have sometimes been made by a company but there is apparently no obligation to do so.

The most noteworthy thing about air-accident compensation is that it is limited to a certain maximum amount, by arrangement among the various international airline bodies

[19] Lloyd's of London, Daily Reports, September 29, 1968.

and governments. But differing opinions about what the figure should be have led to a confused situation.

The Warsaw Convention of 1929 limits the liability of an air carrier, in the case of injury or death to one of its passengers, to approximately $8,300. Note that this is a *maximum*. A court will base its estimate of what could actually be paid on the circumstances of the accident and the passenger's "life-value."

As some consolatory balance for this limit, the Warsaw agreement also laid down a rule which means, in effect, that a passenger does not have to prove the airline guilty of negligence in order to get his compensation. The carrier has to prove itself innocent. The wording says that "the carrier is liable if the accident . . . took place on board the aircraft or in the course of any of the operations of embarking or disembarking," unless he can prove "that he and his agents have taken all necessary measures to avoid the damage or that it is impossible for him or them to take such measures."

This is especially important for a passenger's dependents in those cases where not even a probable cause can be given for an accident: for example, when an aircraft totally disappears, as did a Tudor of British South American Airways on January 17, 1949. Its last known position was about two hundred miles southwest of Bermuda. Twenty lives have never been accounted for in this accident.[20] Or in a crash such as that of a Canadian Pacific DC 4 that disappeared off the face of the earth on July 21, 1951, somewhere around Yakutat in Alaska. None of the wreckage or the thirty-seven bodies has been found.[21]

The cash limit does not apply in cases of "willful misconduct." But this is almost impossible to prove, unless there is some unmistakably deliberate act, like the pilot trying

[20] I.C.A.O. Information.
[21] I.C.A.O. Accident Digest No. 3, p. 49.

to kill himself. The convention terms do not apply on domestic routes—only on international ones—and traffic to a country which did not sign the Warsaw agreement would not be covered.

A further agreement known as The Hague Protocol of 1955, sponsored by I.C.A.O., tried to bring Warsaw up to date. So it partly did; but it also left behind several curious contradictions. The Hague doubled the liability limit to $16,600. (Again this is a maximum; it does not include legal costs—which could take a substantial bite out of this figure if even a few days' court argument were needed.) The phrase "willful misconduct" is replaced by the fractionally broader phrase, "an act or omission . . . done with intent to cause damage or recklessly and with knowledge that damage would probably result." This still appears to leave even gross negligence out of consideration, since the air-accident records show that intent and recklessness are seldom involved, in the ordinary meaning of the words; and knowledge of probable damage would be hard to prove. The phrase is almost irrelevant to the nature of poor safety standards, which have more to do with general neglect and complacency.

The United States, in The Hague talks, tried to have the limit raised far beyond the $16,600 figure, which was not much more than a year's earnings for many of their nationals traveling on international flights. It seemed that the Americans had in mind a figure five or six times as high; and this was unacceptable to many of the small states. Finally it was agreed by all signatories that a limit of $75,000 would be acceptable for flights into, out of, or through the United States. Citizenship is not important; it is the wording on the ticket that counts. If a passenger flies from Montreal to London direct the maximum compensation payable is $16,600. If the flight goes via New York the figure could well be $75,000.

Despite this slight legal backing for air-safety enforcement, the companies have very occasionally had to pay out large sums to dependents. After the Boeing 707 crash at Mount Fuji, Japan, in 1966, 113 claims for compensation were filed against the two companies, B.O.A.C. (the operators) and Boeing (the makers). A Chicago attorney, John J. Kennelly, secured an onlooker's film of the disaster and it suggested the mid-air disintegration of the tail of the aircraft. According to *Newsweek* magazine (July 29, 1968), Kennelly proposed using this evidence to support charges that B.O.A.C.'s own records indicated that the company knew of a defect in the tail section of the aircraft concerned but did nothing about it, and that manufacturing processes had caused structural weaknesses in the tail. Neither charge was proved, *Newsweek* added, but the two companies apparently chose not to argue with Kennelly in court. Kennelly made the important point that the Warsaw limit did not apply in this case, because of various technicalities, one being the tiny size of the print on tickets informing passengers of the airline's liability.

The outcome was the largest known out-of-court settlement in the history of aviation disasters up to that time. The settlement of one group of fifty-four claims was understood to have totaled $8,282,786, an average of over $150,000 per claim. A B.O.A.C. official has since said that they had not contested the matter in court, "because of the costs involved." I queried him about paying out eight million dollars in these claims, and he reiterated that they had estimated that it would be cheaper to pay out this total than to contest the cases in the American courts. And even later news: A B.O.A.C. official has since said, "Both B.O.A.C. and Boeing firmly believe that there was no structural defect which caused or contributed in any way to the accident. It is significant that the claimants also chose not to pursue their allegations in court. The settlements made were certainly well in excess of the Convention limits, but this is not significant because the manufacturer's potential liability was also released by the settlements."

Not all relatives of people who died in this accident received $150,000 (approximately). A man whose mother, father and brother died on Mount Fuji told me he received far less.

As aircraft get bigger, the provision of insurance coverage becomes more problematical. The air frame and third party will be so huge that insurance companies, including Lloyd's of London, doubt very much whether they could also take on passenger insurance. Imagine 900 checks, each of $75,000, just to the passengers.

Specialists in aviation insurance believe that the worldwide total of premiums paid by airlines, now $240,000,000, will have to double by 1975. The two prototype Concordes are expected to be insured jointly for $48,000,000—the cost of around forty of today's lower-priced, older aircraft.

The aviation-insurance stake which will be at risk in the future, even working on Warsaw or The Hague limits, will be a prodigious one. The insurance market is working on the assumption that some four hundred Boeing 747s, each with a capacity of nearly four hundred people, will be operating by 1975. One insurance expert has calculated that if two of these collided in mid-air over the United States, the possible claims for hull, passengers and crew might total $200 million.

The lesson is clear enough. Airlines' insurance premiums went soaring after the Comet crashes, when it became apparent that the safety of the bigger aircraft was still in doubt. Must the same thing happen again? When one still finds the airlines raising arguments against the "necessity" of carrying life rafts on oceanic routes, because of the so-called weight penalty, one is obliged to doubt whether their attitudes are yet adequate to the coming era. Looking purely at the financial aspects, it is in the airlines' interest to see that a better standard of preventive measures is set afoot *now*.

6 · The Charter Game

When we come down the scale to the smaller airlines and the ones which operate charter services, we see how safety margins can be pared down and how intermittent is the control exercised by governments. There is nothing intrinsically wrong with charter flying. Nearly all the big operators take part in it to some extent, hiring out a crewed-up aircraft to a travel agent or some other carrier for one or more trips, usually for a package vacation tour. On these, the passenger can expect to board a plane that is no different from the one he would find on the scheduled services.

But the small charter airline is in a different position. Its capital may not be great and it does not have the steady revenue of regularly scheduled services. In the summer vacation season its resources are fully occupied with cut-rate tourist trips. It can afford the cheap fares largely because it is guaranteed a full planeload.

It would, of course, be far too sweeping to say that all charter airlines are inadequate. Some of them achieve high standards of operation. As the airline safety table in Chapter 5 shows, the small airline that cares enough about safety can establish a minimal accident rate better than a number of the big operators, despite its more slender resources. This in itself is a revealing point.

But the accident figures also show that the safety record of much of the charter business is below what it should be and that the record includes some abysmal cases of negligence. Possibly the first official inquiry into it was that set up by the British Board of Trade in 1967.[1] It found that

[1] "The Safety Performance of United Kingdom Airline Operators."

British charter aircraft had been involved in crashes *two and a half times* more often than those on scheduled services. This situation must have been existing for years without anyone in authority investigating it. The Board of Trade was prodded into making an inquiry only by the public concern aroused by a number of accidents involving aging aircraft of the DC 4 type on charter services. There being no reason to think that Britain is unique in this respect, it is very possible that a similar higher proportion of accidents among charter companies is the case in other countries.

One can see a number of reasons why the charter airlines carry a higher risk. Not one of these factors is absolutely unavoidable. It is also true that, taken one by one, they may not inevitably produce a series of accidents. It is a fact, however, that it takes a poor airline to expose the shortcomings in aviation safety. It is worth reiterating that a company which trims down safety margins in just one sector can often get away with it—whether it is meager equipment, the carrying of minimum fuel supply and maximum payload, the overworking of pilots, or the using of substandard airports. It is when several of these combine that the tragic, so-called "accident" inevitably occurs.

Most of these flights are at night, so that they get the advantage of lower airport charges and a steady, round-the-clock employment of their equipment. They are often flying to the second-rate airports which happen to be closest to the tourists' destination—places like Perpignan, in southwestern France, whose combination of venomous terrain and slight navigational aids had been the subject of frequent complaint in pilots' reports well before a series of crashes made it notorious.[2]

In addition, the charter pilot is expected to take a plane into unfamiliar airports far more frequently than the sched-

[2] IFALPA Annual Conferences.

uled pilot. While the latter usually has a mental picture of the runway lights of Paris, Rome, London, New York, Tokyo, Nairobi, or his other frequent ports of call, the charter man will more often be depending solely on his charts.

Neither night flying nor unfamiliarity nor second-rate airports, individually, need cause great alarm to the experienced pilot. But, put together, they can be disastrous.

Now we come to the aircraft itself. Some companies which specialize in charter are operating older, out-of-date planes which have been pensioned off by the scheduled airlines after long service, or which have come on the market from military and other sources. The Douglas DC 3, a 28-seater, is one of the most common. The first of the type came into service in 1936, and 10,745 were manufactured. There are at least one thousand of these Dakotas in service at present. The sales columns of the aviation magazines still regularly advertise them at a price of around $20,000.

They were very useful aircraft in their day; but the cheapness, which makes them attractive to the company with limited capital, tells its own story. Their original cost will have been written off several times. There are those who argue that the fact that a plane is old, that it may have been in continuous service for up to thirty years, matters little in practice. They point out that to keep its certificate of airworthiness an aircraft must be shown to have had its engines and other vital parts renewed or strengthened at regular intervals. This is a specious argument; it ignores all the bitter lessons we have learned, and have still to learn, about metal fatigue and turbulence in recent years. The crashes of the first Comet are an example.

To restore these old aircraft to what is *presumed* to be a standard fit for commercial service, the manufacturers sell "recertificating" kits. These are used to strengthen parts of the fuselage which have been subject to stress over a certain number of flying hours. We are asked to suppose that this restores the aircraft to a safe condition. But our knowledge of metal fatigue should be enough to warn us to be skeptical about this. The regulations, for instance, do not take into account the fact that vital parts like the wings and wing roots, the tail and fins, have been subject to years of stress. Even minute stresses applied constantly over a period cause a weakening. Under all normal circumstances, no doubt, the plane still flies well. But no one can really be sure how far the risk is magnified when the aircraft gets a severe buffeting in storm turbulence.

These old aircraft have been involved in a large number of accidents over their lifetimes, and in a number of cases it has been left in doubt whether the instability of the plane's structure was not a contributory cause. Nor can we be sure —since the planes have often gone from one operator to another—just how efficiently the maintenance routines have been carried out. So much has to be *presumed*.

There is no more reassurance to be had from considering the plane's flight-deck equipment. Obviously it is out of date, mostly belonging to the piston-engine era; and, because of the original design of the cockpit, there is seldom room to install such devices as weather radar, multiple radio systems, and other aids. The older Viscounts, for instance, lack essential fail-safe equipment—elementary equipment—and this lack has been cited as a contributory cause of a number of accidents.[3]

The British Board of Trade report concluded that *over 30 percent* of accidents involving independent operators were attributable to failures of airworthiness. The most frequent causes were collapse of the landing gear or engine failure. The report specifically points out that these defects are associated with older aircraft. Yet in most countries this sys-

---

[3] See Chapter 7.

tem is allowed to continue. We know the facts, but nothing radical is done about it.

So, for our main line of defense against unsafe standards among the charter operators, we are largely thrown back on the strength or weakness of a government's regulations and the quality of its inspection systems. For instance, one British charter airline is packing them in—and we mean exactly that; a galley and a lavatory are removed (or not installed), and the seats are smaller. The government body that is responsible sanctions this completely as long as the operator can evacuate the aircraft, under the regulations laid down, within ninety seconds.

With the big operators, who are so much more in the public eye and have a greater reputation at stake, it is fair to say that they mostly conform to the inadequate standards of the time. With many of the smaller airlines, however, it has become habitual to operate right up to the limits set by the rules; consequently, there are numerous examples of an operator or his flight crew going beyond them too.

The accident reports illustrate the sort of commercial pressure that can push safety standards down to danger level. All the following examples are taken from the I.C.A.O. aircraft accident digests.

1. A DC 3 of General Airways crashed near Kerrville, Texas, on February 1, 1959, when carrying a full complement of passengers.[4] The captain, the reserve captain, and one passenger were killed; four others were seriously injured. The official inquiry found, among other things, that the plane was overloaded by 517 pounds at takeoff. The weather forecast on its route indicated low ceilings, restricted visibility, snow, freezing precipitation, and icing conditions. The official report says that the crew had been

[4] I.C.A.O. Accident Digest No. 11, p. 79.

on continuous duty for more than forty hours, which was a violation of the civil-aviation regulations. The plane got into difficulties when the wings began to ice up, with its fuel running out, and it probably crashed in attempting a belly-landing.

Evidence given to the inquiry revealed that a pilot was required by this company to act as its agent while away from base. If he had decided to make an overnight stop and wait for better weather, the Company would have had to pay for board and lodging overnight for the twenty-five passengers. According to the I.C.A.O. Accident Digest, the inquiry found that "the action of the Captain in getting so far into a bad situation could be attributable to indifference to elementary rules of flight safety, coupled with severe economic compulsion." The I.C.A.O. Digest stated that the F.A.A. imposed civil penalties on General Airways for its violation of regulations pertaining to flight crew's duty time and takeoff weight; but its operating certificate was restored.

2. A Curtiss-Wright Super C-46F operated by Arctic-Pacific, Inc., crashed shortly after takeoff from Toledo Express Airport, Ohio, on October 29, 1960.[5] It was carrying forty-five passengers and three crew. Twenty passengers and the captain and copilot were fatally injured. The inquiry found, among other things, that the captain had attempted a takeoff in zero visibility with the sky nine-tenths obscured by fog. He experienced loss of power on one engine during the takeoff run, made a premature lift-off, and crashed about a mile beyond the runway threshold. The plane was, among other things, found to have been 2,009 pounds over its maximum permitted takeoff weight for this airport. The inquiry was told that the pilot would have been responsible for the subsistence expenses of all his pas-

[5] I.C.A.O. Accident Digest No. 12, p. 299.

sengers if he had delayed departure. The F.A.A. suspended Arctic-Pacific's operating certificate, and it was not renewed.

Comment on these two cases is scarcely necessary. The F.A.A. cannot be regarded as blame-free. It must have known of the system by which pilots acted as agents and so had a vested interest in getting a flight moving, whatever the conditions. To allow such a conflict with basic safety standards seems unpardonable.

Other examples illustrate deficiencies to which charter companies seem prone. Poor engine maintenance is one of them—a doubly risky thing when combined with illicit overloading. Pilot fatigue, as a consequence of working excessive hours, usually for a bonus payment, often occurs. So do instances of slack professional standards of safety among flight crews and ground staff. These are not isolated matters of chance: they almost invariably turn out to be a reflection of inefficiency in the organization as a whole, from top management down.

3. A Catalina belonging to Asociación Interamericana—a company operating in Colombia, South America—crashed soon after takeoff from Bogotá on December 8, 1956.[6] Fourteen people were killed and two were seriously injured. According to the I.C.A.O. Accident Digest, the board of inquiry found, among other things, that the plane was overloaded to the extent of 1,505 kilograms (about 3,300 pounds) over the recommended operational weight for Bogotá; that "the company had not established an adequate technical system for overloading"; and that the aircraft's dispatcher "lacked the technical knowledge necessary for the proper discharge of his function." The aircraft had not undergone the necessary flight test after the replacement of a turbine section of its starboard engine. The starboard

[6] I.C.A.O. Accident Digest No. 8, p. 166.

engine had been repaired hastily and without conforming to technical standards a few hours before the accident. There was evidence that both the pilot and the flight engineer were suffering from fatigue due to overwork at the time of the crash.

4. A Lockheed Constellation of Imperial Airlines (a charter company operating in the United States) crashed on November 8, 1961, while attempting to land at Byrd Field, Richmond, Virginia, with three engines out of action.[7] Seventy-seven people were killed; two escaped with injuries.

The Civil Aeronautics Board accident inquiry says that "this flight crew was not capable of performing the functions or assuming the responsibility for the job they presumed to do." After an intensive inspection by the F.A.A., "numerous improper operational procedures and maintenance practices" were found. "It is believed that the substandard maintenance practices of the company's employees were condoned by the management," the report said.

5. A Lockheed Super Constellation belonging to the Flying Tigers Line (a United States charter company) was on a trans-Atlantic flight to the United States on September 23, 1962, when two of its four engines began to give trouble.[8] The pilot could not maintain height, so he decided to "ditch" in the ocean. Plenty of time was available to prepare for this, as it was not an immediate emergency. Seventy-six people were aboard the plane.

The stewardesses, the board of inquiry was told, briefed the passengers on ditching procedures. As there were differences between these instructions and those given in the safety booklets in the seat pockets, some passengers became

---

[7] I.C.A.O. Accident Digest No. 13, p. 251.
[8] I.C.A.O. Accident Digest No. 14, Vol. 1, p. 60.

confused. When the aircraft ditched—without any signal having been given to "brace" at the moment of contact—a number of passengers were in an upright position instead of in the proper crouch. Of the five twenty-five-man life rafts carried aboard only one was brought into use, and fifty-one persons boarded this raft. Twenty-eight people lost their lives in this accident.

6. A Boeing 307 of the French charter company, Compagnie Air Nautic, was on a flight from Bastia to Nice on December 29, 1962, with twenty-five people aboard.[9] The plane flew into a cliff on the island of Corsica and then fell 300 feet to the beach. All aboard were killed.

The inquiry found that though the crew had not exceeded the authorized flying hours, they probably had not had a complete night's rest since December 26 (three days before). The board concluded in part that the crew had not been given the necessary instructions for the route. Those contained in a copy studied by the board were found to be "inaccurate and dangerous," since they did not have the altitude at the turning point and they provided a flight time between Bastia and Ajaccio which was incompatible with the attainment of safe altitudes.

7. A Viking owned by Independent Air Travel Ltd., a British charter company, was on its way from London to Tel Aviv on September 2, 1958, to pick up a load of passengers.[10] When it crashed three miles from London Airport all three crew members were killed, together with four people on the ground; nine others on the ground were injured, some seriously.

The inquiry revealed in part that maintenance on the plane had been carried out "by tired men, working under

[9] I.C.A.O. Accident Digest No. 14, Vol. 1, p. 108.
[10] I.C.A.O. Accident Digest No. 10, p. 202.

without proper supervision or instruction." Another it found was that the pilot had been on duty for and a half hours—which was described as a "gross breach" of the rules, since the maximum permissible flying-duty period for a two-pilot crew was sixteen hours.

"The gravity of this matter and of the disregard of the regulations in the case of the rest time to which the captain was entitled became apparent when it was disclosed that the company had been prosecuted and convicted in May 1958 on ten charges involving breaches of the regulations governing flight-time limitations and that these convictions involved both excess hours and insufficient rest periods accorded to pilots." The report stated that "the Directors of the company put the whole blame for the accident on the Captain."

But the Board said that "the conduct of the pilot and the whole course of events outlined were contributed to by the deliberate policy of the company, which was to keep its aircraft in the air and gainfully employed, regardless of the regulations for the conditions of working of its employees or the maintenance of its aircraft."

It added that since the accident the company had taken great pains and spent a good deal of money in putting its affairs in order, with the result that its organization now (in 1959) bore favorable comparison with that of the larger companies. It folded soon after, but this case suggests laxness on the part of the Board of Trade. It fined this charter company for ten breaches of regulations within a short period of time—and permitted it to continue operations, until finally innocent people were killed.

It should be impossible to find such examples of malpractice and inefficiency in *accident* reports. If an investigation of one crash reveals that a pilot has been grossly overworked or that engines have been badly maintained then it could also suggest that these practices have been habitual in the company concerned. When government inspection is supposed to exist, why have these malpractices not been spotted long before?

There could be several answers to this. One might be that the standards required of an airline before it gets an operating certificate are too low to begin with. This is unwittingly revealed in a paragraph in the Board of Trade report.[11] "The Director of Aviation Safety," it says, "has so far granted certificates to eighty-eight operators; twenty-six have lapsed for various reasons; and twenty have been suspended or revoked. On nine occasions, when operators' standards were seen by the Inspectors to have fallen below an acceptable level, the Director has given notice of his intention to suspend or revoke unless an improvement was effected; in six of these cases the Director was able to withdraw the notice upon being satisfied that the necessary action had been taken by the operator."

The passage is worded in a way that suggests the alertness of the director and his inspectors. This may be true, up to a point. What is really remarkable, but is given no special emphasis, is the number of companies which have been given permission to carry passengers and cargo and subsequently found to be unfitted for it. This happened in twenty cases, or nearly a quarter of the total. Adding those which received the threat of suspension and some of those companies which let their certificates lapse, one finds a frighteningly high proportion of airlines that have been allowed to operate for a period of time on apparently shoestring standards. If this has been the case in Britain, can we be sure that it is not happening in every country? All pay lip service to safety. Many, judging by the accident reports, allow incredibly low standards of efficiency in some sectors of the industry. These charter airlines are allowed to advertise their cheaper fares. Obviously, there is no requirement to publish the fact that

[11] "The Safety Performance of United Kingdom Airline Operators."

statistically their over-all accident rate is two and a half times that of the scheduled operators. Until a series of crashes makes it blatant and unavoidable, there appears to be no inclination to investigate the reasons for this high figure.

More surprising facts emerged from this British inquiry. The Board of Trade is ready to grant operators' certificates, although it knows that it does not have an inspectorate adequate for the job. The figures I have quoted show how frequently the standards of an airline prove to be faulty once it has got its certificate and is in business. The Board of Trade report notes, in addition: "It is evident that there are some who meet the minimum standards evolved for the Air Operator's Certificate only by virtue of the close supervision and constant prodding to which they are rightly subjected."

Note that they have to be constantly prodded to meet even *minimum* standards. It is staggering to find that any authority which publicly professes concern about air safety allows such lackadaisical companies to stay in business. For while this official "prodding" is going on, passengers are still being carried and deceived about the degree of risk.

Nor can we be sure, at least in the British case, that the inspectorate have been able to discover more than a portion of the operational shortcomings of certified airlines. As the report puts it: "The inspectorate has been consistently short of staff; only during the last few months has the effective number of Inspectors risen above half the complement that the Treasury now accepts as necessary." If a group of inspectors is allowed to operate at less than half strength—and even then finds so many operators below the minimum standard—one can only wonder what a properly manned body would uncover. It has been too often the case that a crash has been needed to disclose a glaring disregard of safety, whether or not an inspectorate exists. There is, in one of the examples of accidents mentioned above, the case of an airline which proved to be overworking pilots only four months after it had been prosecuted for a similar offense. It is clear that weak supervision encourages contempt for the regulations.

The report rightly recommends "a much more intensive programme of inspections in flight," since this is the only way in which the tendency to overwork a flight crew can be countered. What else can be done? The standards needed for an Air Operating Certificate need to be updated in all countries. Pilots' working hours—to mention only one aspect of the matter—are still based on the flying schedules of the 1950s, whereas every passenger who has made a long flight through a number of time zones knows how disruptive jet travel can be to his rest pattern.

Those authorities who grant certificates must be more strict in their standards, particularly where they know that their thoroughness of inspection is not up to scratch. If a company can operate only on *minimum* standards, then it is plainly not good enough, since today's minimum standards are themselves on the verge of being hopelessly out of date. It should stay out of business. Flying itself can ruthlessly expose the inadequate. So it is better that the licensing authority operate in a similar manner *before* the crash occurs.

7 · Designed for Disaster

Not one civil aircraft flying anywhere in the world today is designed to be as safe as it could be, on the basis of present knowledge and a common-sense appreciation of past experience.

In a number of other respects the performance of several modern aircraft types can fairly be described as excellent, whether in speed, the smoothness of the ride in normal conditions, or in those few other appealing characteristics that figure in the advertisements.

But if we carry our inquiry back down the line to the factory floor and take a critical look at the ideas that are built into any new aircraft (besides noting the ones that are so regularly omitted) then it becomes plain that passenger safety is left to be the poor relation of the system, and that this is the case from the earliest stages of conception.

The development of a new aircraft is naturally a complex, many-sided process. Fundamentally, it rests upon an interchange between the airlines, with their various traffic requirements, the manufacturer with his marketing objectives, and the civil-aviation authority of the country concerned, which lays down the minimal-performance criteria. It is thus hard to isolate any one of the three and to demonstrate that it is more blameworthy for low safety standards than the others. It is certainly within the power of any one of them to effect an improvement. Insofar as he has the ultimate say about what goes into a design, and on the degree of priority that will be given to safety in his general research and production effort, then in many respects the manufacturer must take a major share of the responsibility for the inadequate standards.

Of course, no sane manufacturer would put an unsafe aircraft on the market deliberately or even, I must suppose, if he had any overwhelmingly conscious doubts about its capabilities. The aircraft must, in any case, conform to certain minimum standards. One must accept that a manufacturer gives a good deal of scrutiny to the safety qualities of his products—but, I would add, only applying the standards considered "normal" by the aviation industry and its various supervisory agencies at the time.

Even so, there have been cases where major errors of manufacture or design have slipped through the net. The De Havilland Company's first Comet and, later, the Lockheed Company's Electra caused loss of life as well as heavy financial loss to the companies. In these cases one has to assume that the manufacturers had a great deal of research experience behind them, working to standards accepted as "normal" within the industry.

It could be reasonably argued that the industry lays itself open to such occasional and unusual disasters simply because it sets its own sights so low, so close to the *minimal* requirements of the government authority. In aviation no incident is truly isolated. The Comet and Electra crashes were symptoms of the malaise of the whole system.

It is true that one can delve into the accident records, examine some hundreds of reports by boards of inquiry, and still find *relatively* few in which the investigators have put the blame firmly on some defect in design or manufacture. In severe crashes, it is extremely difficult for the investigators to tell with certainty what caused the crash, since much of the evidence is destroyed. So their findings are normally only tentative. Boards of inquiry are invariably reluctant to pin the blame firmly on any one agency, partly because in most accidents it is difficult to apportion the blame among the various parties involved. The ways in which any particular aircraft design falls short of the truly safe can also be a subtly difficult thing to establish.

Another problem is that the investigating agencies, fearing legal or political reprisals, are often reluctant to directly accuse the responsible parties. This is especially true when an aircraft crashes "on foreign soil." The position could be corrected by I.C.A.O. taking charge of all accident inquiries and changing world civil-aviation laws to give legal immunity to crash investigators and to require the investigators to name publicly the products, person, employees and companies responsible for the crash and subsequent loss of life. If this were done, one would notice that the conclusions of the crash reports would start: "The causes of this crash were as follows . . ." instead of, "The *probable* causes of this accident are as follows . . ." with which crash reports now usually preface their timid conclusions.

Some agencies thrive on this. By advancing their conclusions merely as probabilities the investigators provide an easy way out for the company, person or airline involved, and also for their own agency, which, in a large number of cases, is also responsible for formulating and administering the civil-aviation regulations broken by a company, airline, or persons because of the agency's inadequate policing.

Unfortunately there are numerous cases that I know of on very good authority, that could only be described as complete dishonesty and withholding of facts. The most recent case happened a couple of years ago in a crash in which more than forty people were killed or injured. The information that I have, which was not made public at the crash inquiry, would turn hundreds of people away from the airline concerned; but it is a nationalized airline and thus is closely allied with its sister government department that investigates the crashes.

If, for example, a certain design of aircraft shows a tendency to lose height quickly once speed is reduced for landing—that is, to have a high "sink rate"—then it is difficult, if not impossible, for a board of inquiry to condemn the manufacturer concerned. It would have to bear in mind that other flight crews have landed the same model successfully thousands of times; it would have to postulate for comparison some safer design that would not have run into trouble. So the verdict goes against the personnel as "pilot error." No mention is made of the objectives that led to the production of an aircraft of this sort.

Safety is indivisible. It is a chain in which such apparently dissimilar aspects as pilot skill, well-placed emergency exits, good flying instruments and strong seats are all linked. And we can only see the connection between them once an accident has taken place—when shortcomings in two or three or more of these things happen to coincide.

But a large part of the trouble is that the industry persists in seeing itself as departmentalized, split off into disconnected parts without any honest relationship with one another where safety is concerned. This seems to be one of the illusions of which the manufacturers are most guilty. Each sector of the industry persistently unloads part of its own share of the responsibility for safety onto the next sector, admittedly most of the time being blandly unaware of what it is doing.

The airport authority presupposes that there will be pilots who are always in top form and aircraft that can always cope with its rather badly laid-out runway. The manufacturer presupposes entirely adequate runways and, though he will add a margin to the braking power in case of error, it strangely turns out not to have been adequate in this or that accident, when the pilot's reactions were not at their best and when he had a tricky cross wind to deal with too.

Accidents eventuate, as I have suggested, when several

deficiencies happen to coincide. To assert that because some operators fly a variety of aircraft types safely, thus giving the manufacturers concerned a clean bill of health, is to get the whole picture wrong and divert us from the truth. It means that the airlines concerned have set such good standards in other departments of their operation that they manage to neutralize this tendency of accident factors to occur all together at some point in time. If, instead of looking at the safest airlines, we examine the record of those that have had accidents, we can see how often the designers have contributed their element of weakness to the chain. It must also be remembered that even the safest airlines have a daily list of unpublicized "incidents" to report—that is, occasions when the flight crew have registered one or two things going wrong and when it only needs the third and fourth elements, or the fifth and sixth, to push them into the accident area.

Since aviation design is a highly technical subject it is also necessary to question the tendency of laymen to accept the designers' products as somehow unquestionable. This would be a poor start to an investigation; so let us look at the Lockheed Electra, the brain child of a normally skilled, conscientious and well-meaning design team.

On September 29, 1959, a Braniff Electra was flying from Houston to New York International when, in a clear moonlit night at nine minutes past eleven, the left wing snapped off, probably following a short period of undampened propeller whirl mode. This means that the No. 1 (outboard left) engine nacelle and prop probably started "bouncing." This progressed until it was so severe that the left wing, left landing gear, and the No. 1 and No. 2 engines snapped off from the fuselage near the wing root. Thirty-four people died when the fuselage, tail, and starboard wing crashed to the ground in flames.

N9705C was a new aircraft. Its final assembly was started in April 1959, and the first of its three production test flights was on September 4, 1959—twenty-five days before the accident. Braniff received delivery of the Electra N9705C at the Lockheed plant on September 18, 1959. Braniff reported that the only chronic difficulties with the plane were with the radio navigational equipment and the generator, which had malfunctioned during the preceding few flights. None of these faults had anything to do with the No. 1 engine whirl mode oscillations.

The Lockheed Corporation lost $118,000,000 when they had to call in all the Electras that were in service. The inspectors who saw them as they came in were surprised that these aircraft had been able to keep their wings in flight. After the first crash those who knew about the design exerted pressure to have the aircraft withdrawn, but this was resisted. One of the first people to realize that something was radically wrong was Edward J. Slattery, of the National Transportation Safety Board.

The upshot was that Lockheed had to pay to have the wings and wing roots of all the Electras in operation strengthened, and most of their factory space had to be converted to make these changes. Edward Slattery himself told me: "One condition of being allowed to go on flying them was that they not be flown at more than half of their normal cruising speed. But once we saw that their wings were staying on more by luck than by design we realized that a potential danger was present even when taxiing these things on the airport."

Yet this aircraft must have passed enough tests during its development to satisfy the manufacturer. As with every other new type, it had been built under the constant supervision of officials of the Federal Aviation Agency. Exactly how this weakness in design escaped detection is obscure. All that can be said is that it raises serious doubts about the infallibility of the system of testing and certificating aircraft. But in this case the sickening thing is that this outer propeller oscillation made itself fatally evident after only 132 hours of flight—or eleven days' service. Surely the initial testing program should have amply covered this period of existence. Since the Electra disasters, Lockheed has been right out of the civil airplane market, only now reentering it with their Model 1011, the Tri-jet.

The failure of the first Comet was even more puzzling. Two aircraft of this type were lost within a period of eleven weeks in 1954, having apparently broken up in flight at about the time at which they would have reached cruising altitude.[1] It took a long and immensely expensive investigation to establish that the cause had been metal fatigue. The scientists at Farnborough Research Station had to piece together thousands of fragments recovered from the sea bed, while another Comet, used as a guinea pig, had to be put through some thousands of hours of simulated flying time in a water-tank test before the answer was found. It appeared that a crack could suddenly emerge in the fuselage after the cabin had undergone the pressurization routine a certain number of times—with, inevitably, explosive effect.

Each year, a number of deaths are shown to have been largely attributable to some basic flaw in the structure of an aircraft; something which could and should have been remedied on the factory floor before the plane was released for service. There is a somewhat larger number of cases in which this kind of defect has been adjudged a merely contributory element in an accident. Nor should we forget the number of occasions in which some malfunction or handling difficulty is mentioned in every airline's daily "incident" log. Normally these are harmless, because the manufacturer's own fail-safe devices have worked properly or the flight crew have been able to take prompt action. But they could, in many circumstances, become contributing factors to an accident.

[1] I.C.A.O. Accident Digest No. 6, p. 16.

Once these shortcomings are noticed they can usually be put right throughout the whole fleet, or else pilots can be warned to take account of them. But, in studying the accident reports, I am disturbed to notice how often it has required an accident, involving death and injury to passengers and crew, to reveal a defect that should have become obvious in the factory testing phase. A classic example of this was the Argonaut crash in Stockport, England, in 1967, already mentioned in Chapter 1. One accident disclosed that the engines of an aircraft that had been in service for twenty years could be starved of fuel without the pilot being aware of the hazard.[2]

The designer's job embraces a wide variety of factors and problems: the aerodynamic or "handling" qualities of the aircraft; the positioning and number of engines; the strength-and-weight ratio of the fuselage; the layout of the cockpit; the location of emergency exits; the performance of instruments and engine controls; the correct placing of fuel lines, and so on. No one can pretend that they present an easy task, particularly when it is often a matter of striking a balance between the varying requirements of different customers.

One has to accept the elusiveness of perfection. That said, what should arouse concern is the way in which, from the point of view of passenger safety, errors of design and shortcomings in equipment are perpetuated from one type of aircraft to its successor, even though airline "incident" logs and accident reports will have challenged their good sense on numerous occasions. One is obliged to wonder whether the manufacturers study these reports, or whether they know about these deficiencies and regard them philosophically as part of the "normal" risks of flying.

Even the celebrated Boeing 707, which has probably carried more passengers than any other type of aircraft, and

[2] United Kingdom Board of Trade Accident Report No. **CAP 302**.

has by now been well and truly tested on the world's air routes, was not without its initial problems. As soon as it was in service the rudder gave trouble. There were three accidents between 1959 and 1961 (in which nine people died and 14 were injured) probably because of problems with the rudder that gave the pilot the impression that it was jammed.[3] Modifications were made to it, also to the tail fin, because of aerodynamical problems. Two more crashes in 1962 (225 dead, 2 injured) were probably due either wholly or partly to trouble with the rudder and stabilizer trim servo motors.[4] The Boeing 707 like most other aircraft is said still to possess some special handling problems, of which a pilot has to be wary. An accident investigator has said that were a pilot to allow the 707 to get into a 35-degree nose-down angle of flight it would be very difficult indeed for him to pull it out of a dive.

Another four-engined jet in widespread service around the world, the Douglas DC 8, has run into a number of problems, some of them serious.

A crash in 1961 (18 dead, 12 injured) revealed, among other faults, that there was no dual hydraulic system for either the wing-spoilers or the tail fin and rudder.[5] A fault in the hydraulic system was cited as the probable cause of the accident causing two engines not to go into reverse thrust when it was selected. This is a flaw which is nearly comparable to owning a Rolls-Royce or a Cadillac without a handbrake.

These wing-spoilers, which you see from your cabin window, are the flaps that operate as the aircraft begins its descent. They stick up from the top surface of the wing. By interrupting the flow of air over the wings they reduce the

[3] I.C.A.O. Accident Digest No. 11, p. 190; No. 12, p. 35; No. 13, p. 143.
[4] I.C.A.O. Accident Digest No. 14, Vol. 2, pp. 22, 71.
[5] I.C.A.O. Accident Digest No. 13, p. 123.

speed. If the hydraulic system breaks down, it is important that the flight officers know of the seriousness of the fault and take remedial action. A dual system would halve the probability of a complete loss of wing-spoilers. Some civil-aviation regulations specifically state that this dual capability must be available for emergencies. Despite this omission, for some inexplicable reason the government aviation officers had permitted an airworthiness certificate.

An inquiry into another accident involving a DC 8, in November 1963 (at Montreal; 118 dead),[6] criticized the vertical gyro warning system (which tells the pilot if his aircraft's attitude is incorrect). It was added that there was need for an instrument that would show the horizontal stabilizer position of the aircraft. In a further accident, in February 1964 (at New Orleans; 58 dead) the possibility of misrigging the pitch-trim compensator on the DC 8 came in for criticism from a board of inquiry.[7] The list of curious doubts about basic design qualities did not end there. A training-flight accident in July 1966 (2 dead, 3 injured), showed that the pilot of a DC 8 can, in rapidly pulling back the throttles, inadvertently put his engines into reverse thrust—that is, cause the four jets to blow air forward instead of backward, in a braking action of great power.[8] In normal circumstances, in most aircraft, the only occasion on which it is used is when the pilot wants to pull up after a safe landing. The idea of reverse thrust coming into play inadvertently in mid-air or, as in this case, during a takeoff, is a nightmarish thought for any flight crew. It is hard to believe that a modern jet, made by a famous and experienced company, can suddenly prove to have this flaw after so many millions of miles of successful flying. The Douglas DC 6, a predecessor of the DC 8, had had problems which might have

[6] I.C.A.O. Accident Digest No. 15, Vol. II, p. 68.
[7] I.C.A.O. Accident Digest No. 16, Vol. II, p. 39.
[8] I.C.A.O. Accident Digest No. 16, Vol. II, p. 68.

acted as a warning. Accidents to this aircraft in February 1952[9] and April 1955[10] showed this alarming possibility. Thirty-six people were killed and thirty-four injured in these crashes.

The DC 6 was involved in a number of incidents in which fires occurred in flight. Investigators found that probably the aircraft's wing-tank overflow and the intake of the air-conditioning system were close enough for fuel to enter these air ducts under certain circumstances. An accident in February 1957 (20 dead, 78 injured) aroused concern that it was difficult to evacuate the DC 6 after this crash,[11] because at least one main door was jammed.

The DC 7 had a number of troubles. Modifications to the elevator spring tabs (these are part of the mechanism for flap retraction; flaps are most important during takeoff and landing phases of flight) were necessary after the findings of an inquiry into an accident in March 1962, when 111 passengers were killed.[12] This inquiry revealed a possible jamming of these units. More replacement changes were necessary after an accident in December 1955 (48 saved). The probable causes were that the propeller governor drive shafts could fail in flight causing overspeeding, the inability to "feather" the propeller concerned, engine failure and subsequently a fire.[13]

The Caravelle, made by the Sud-Aviation Company, has been the principal contribution of France to the jet age. Many travelers commend it for quiet and comfort. However, its safety standards are open to criticism on a number of points.

Caravelles have several times been in trouble on landing,

[9] I.C.A.O. Accident Digest No. 3, p. 127.
[10] I.C.A.O. Accident Digest No. 7, p. 98.
[11] I.C.A.O. Accident Digest No. 9, p. 45.
[12] I.C.A.O. Accident Digest No. 14, Vol. II, p. 36.
[13] I.C.A.O. Accident Digest No. 7, p. 221.

when the axle of the nose wheel has broken. In one such accident, in March 1964, all on board were uninjured.[14] Another accident, in July 1963, showed that it was possible for the emergency exits to lock tight after the stress of an impact—the safety lock on one door was jammed.[15] There were no casualties in this one; but it is disturbing to find a design in which this can happen. Then both the forward doors were found jammed shut after an accident in Hong Kong Harbor on June 30, 1967, although the impact was classed by the board as "not severe."[16]

Another questionable feature is the way in which the instrument positions on the cockpit panels of the various Caravelle models differ.[17] The accident-investigation committee noted this as the third in a series of similar accidents. This suggests to me a serious lack of imagination on the part of the designers. I cannot believe that superhuman foresight was required to see that, one day, this was going to cause confusion during a critical maneuver. To make it worse, instruments meaning quite different things were similar in their mode of display. In April 1964 a pilot of Middle East Airlines, coming to land at Dhahran, Saudi Arabia, may have taken a reading from the wrong dial, and thus caused a major accident in which forty-nine people were killed.[18]

A number of aircraft designs reveal an unaccountable failure on the part of the plane makers to understand the pressures of the pilot's job, and the elementary necessity of putting no obstacle in the way of quick reactions. It is all the more difficult to comprehend since most designers must have had a good deal of flying experience in the cockpit. The layman is entitled to ask whether they consulted experienced

[14] I.C.A.O. Accident Digest No. 16, Vol. I, p. 109.
[15] I.C.A.O. Accident Digest No. 15, Vol. II, p. 92.
[16] I.C.A.O. Accident Digest No. 17, Vol. II, p. 179.
[17] I.C.A.O. Accident Digest No. 16, Vol. I, p. 151.
[18] I.C.A.O. Accident Digest No. 16, Vol. I, p. 151.

pilots at each stage of the creation of the cockpit layout. But, judging by results anyway, this does not seem to happen. Another dubious feature of the Caravelle—as pointed out by an official board of inquiry—is the pilot's blind areas in the cockpit window.[19] This was noted as a probable factor in a mid-air collision, in May 1960, of a Caravelle and a light aircraft near Orly (Paris) airport. One person died and seven others were injured; for the survivors it was a lucky escape.

The ways in which plane design too often puts an extra burden on the flight crew reinforces one's doubts about that convenient verdict of boards of inquiry, "pilot error."

Next, the Viscount, manufactured by the British Aircraft Corporation, a medium-range aircraft in widespread use around the world. There have been a large number of accidents involving this aircraft, and a couple of years ago there seems to have been an excessively large number of Viscount undercarriage difficulties during landing.[20] Every one of these landing mishaps places the passengers under a hazard of death or injury. Every one ties up busy airports for hours. And quite often they end in damage to the aircraft and the airport. But these incidents and accidents persist with monotonous regularity.

The efficiency of some of the Viscount's components has had to be called into question at a number of boards of inquiry. The official investigation of one crash in November 1957 (training flight; 5 dead), showed electrical-circuit faults in the de-icer system as a probable main factor.[21] (In fairness, the manufacturer does not fully agree with the board.) Examples of other accidents involving the Viscount, with their dates and the probable contributory factors discovered, were: fault in stall-warning device (April 1959, 47

[19] I.C.A.O. Accident Digest No. 12, p. 174.
[20] Lloyds Aircraft Accident Lists.
[21] I.C.A.O. Accident Digest No. 10, p. 45.

177 Designed for Disaster

dead);[22] metal fatigue following a modification, with consequent fracture of bolt securing aileron (March 1957, 20 dead);[23] in-flight fires in cabin air-blower systems (May and October 1964, all persons saved; and September 1966, 24 dead);[24] installation of a badly manufactured trunnion, one of the main mechanical arms controlling the raising and lowering of the nose landing gear (November 1957, all saved);[25] malfunction in the propeller-control switches, resulting in an abrupt loss of lift (February 1956, 12 injured).[26]

And now we come to the crash onto the autobahn south of Munich (August 1968, 48 dead). Preliminary indications given by the accident board of inquiry point to the fact that the plane most probably suffered a major electrical failure in flight, and probably, because there was no auxiliary electric power to operate vital instruments, the pilot lost flight control and/or positioning. The board of accident inquiry is recommending electrical modifications to these aging aircraft.[27]

Referring back to those fires in the cabin air-blower systems mentioned above, Air Canada,[28] after a complete investigation of their fire, which spread to a wheel bay after its initial outbreak during a ground run, wrote to the manufacturers of the Viscount—the British Aircraft Corporation —on June 23, 1964, telling them of the incident and outlining actions taken.[29]

A B.A.C. official wrote back to Air Canada on June 26

[22] I.C.A.O. Accident Digest No. 10, p. 113.
[23] I.C.A.O. Accident Digest No. 9, p. 76.
[24] I.C.A.O. Accident Digest No. 16, Vol. II, p. 73.
[25] I.C.A.O. Accident Digest No. 9, p. 238.
[26] I.C.A.O. Accident Digest No. 8, p. 51.
[27] *Sunday Times* (of London), May 31, 1970.
[28] I.C.A.O. Aircraft Accident Digest No. 16, Vol. 2, p. 73.
[29] Australian D.C.A. Accident Report, October 4, 1967.

saying that the comments "would be passed on internally." A B.A.C. assistant service manager revealed that the matter had been discussed and the conclusion had been reached that no further action was necessary.

B.A.C. didn't even think that the incidents warranted discussion with the parts manufacturer, Godfrey Precision Products. Then on October 5, 1964, British West Indian Airways contacted the British Aircraft Corporation regarding a practically similar occurrence. Only this time the fire had occurred in flight, near Barbados on September 20, 1964. B.A.C. described the incident as "a slight fire" and suggested that the airline look up the appropriate repair manual.

As a direct result of the British Aircraft Corporation's bad reportage methods, bad communications system and a seeming complete lack of genuine service to their customers and their customers' customers, twenty-four people had to die in September 1966, in an "accident" that should never have happened. Following the crash of a McRobertson Miller Airlines' Viscount in the northern part of Western Australia, the Australian Department of Civil Aviation prohibited the use of Viscounts series 700 for the carrying of passengers.

The compactness required in aircraft design makes it difficult to alter or modernize the structure once it has aged in service. This is all the more reason for applying the strictest safety principles throughout the design stage, so that such examples of thoughtlessness as those given above are eradicated from the start. If one looks at a truly veteran aircraft, like the DC 3 (the "Dakota"), one can see how its faulty safety features have persisted throughout its lifetime, though it is noticeable that they have not prevented it from getting a regular award of certificates of airworthiness. It has been involved in numerous accidents; though it is only fair to add that it has done a prodigious amount of flying, often in and out of the more primitive airports. It is a good (or bad) example of what the certifying authorities will accept as safe for passenger transport.

Because the cockpit cannot be enlarged, it is impossible to fit modern safety aids for the pilot. So he often has to fly without weather radar, up-to-date radio equipment, improved altimeters, or automatic-landing gear and, in a number of cases, without ILS systems. In certain models the artificial-horizon instrument is hidden behind the pilot's control stick, making it virtually impossible to see in some circumstances.[30] This instrument is a most important one, since it tells the pilot whether the aircraft's nose is pointing up or down—not otherwise easy to determine quickly at night or if the ground references are otherwise obscured. Such careless layouts can also be found in the latest jets.

The manufacturer who is putting a new design on the market naturally has to keep his basic price down to a level that will match that of his competitors. He must also show his airline customers the best chances of profitable operation. He knows that they will look hard at its fuel consumption, its passenger and cargo capacity, its unladen weight, and so on.

But the truth is that economy and safety are rarely compatible. So all kinds of compromises come in from the start, to determine what is put into an aircraft and, just as important, what is left out of it. Occasionally the little flaws in a design are obvious to the expert eye. But more often they are subtle. Pilots (who do not complain enough, in my view) get into the habit of making do with instruments that are hard to read or windshields that give insufficient all-round view. It is not until an accident happens that a board of inquiry learns, to its surprise, that this or that weakness has existed in an aircraft ever since it left the factory.

But surely, the reader may object, the manufacturer has to observe standards of airworthiness laid down by his gov-

[30] I.C.A.O. Accident Digest No. 16, Vol. 1, p. 91.

ernment? True enough, but these are *minimum* standards, and if one looks closely at the regulations one can see how limited they are in range. For example, a four-engined aircraft must be able to maintain height and speed flying on two engines. It must be capable of reaching a height of fifty feet within a certain distance of takeoff; must be controllable under certain weather conditions at all heights; and must be able to withstand certain pressures when flying through turbulence.

Occasionally, once it has reached the testing stage, it is found that an aircraft does not fullfil one or another of the government's requirements. In this case the controlling body will reexamine the matter and, if it decides that the aircraft's over-all flying qualities are not affected, then the certificate is granted.

While governments appear to keep a close watch on the standards of manufacture, with their inspectors present at the factory during every stage of the creative process, there are reasons for doubting the effectiveness of the system. By the time an aircraft gets to the testing stage perhaps as much as two hundred million dollars may have been spent on its development. The inspectors cannot be wholly detached from the financial aspects. If they want a design alteration that will mean additional expenditure, they have to remember that this, by pricing up the design, may mean the loss of contracts to foreign competitors. The regulations are also negative in tone, where safety is concerned. They simply lay down the bare fundamentals which make civil aviation conceivable at all. There is a wide area that deals with the more positive aspects of safety, and this is left untouched or else the standards set are several years out of date. The regulations give little or no positive impetus to the provision of truly adequate emergency exits, passenger seating, or the most up-to-date cockpit equipment for the pilot. These things would cost money; and it is this false economy that sets the standards.

The T-tail jets—a group that includes the Trident, BAC 111, Boeing 727, Fokker F 28, and the DC 9—are good examples of a type of aircraft that developed out of the industry's insistent drive for economy of operation.[31] The T-tails are in the main (that is, except for the VC 10 and Russia's equivalent) of the short-to-medium-haul range and it is therefore uneconomical to have four engines on them. The designers wanted to produce a "clean" wing—that is, one without anything on it to cause drag, which requires more power and consequently uses more fuel. So they mounted the engines at the rear of the fuselage. With a thin, streamlined, drag-free wing they produced a fast aircraft, which at high altitudes uses about three quarters of the fuel consumed by aircraft with a similar number of engines mounted on the wings.[32]

There were some serious troubles at the start. The T-tails showed themselves capable, under certain conditions, of "deep-" or "super-stalling." Both the Trident and the BAC 111 crashed during their tests. In the case of the Trident accident, it took just two and a half years for details to be made public, even though every facet of the accident was known in the industry a few weeks after the crash. This seems an excessive delay, considering that the public has a right to the fullest information about the causes of accidents and in the shortest possible time.[33] The stall characteristics of the T-tail were known in the middle 1950s, long before the civil aircraft crashes during testing. The discov-

[31] *Aerospace Safety* magazine, U.S. Air Force, May 1965 by Capt. John A. Morrison (republished I.C.A.O. Accident Digest No. 16, Vol. I).
[32] "Aerodynamic Features of a T-tailed Aircraft" (McDonnell Douglas Corporation).
[33] Board of Trade Civil Aviation Report CAP 311.

ery was made during the design tests for a United Kingdom military aircraft.

The Boeing 727 was already in service when it had its so-called "teething troubles." No fewer than four of these Boeings crashed (264 dead, 39 injured, 13 saved), all the crashes in the "let-down" phase (when an aircraft is losing altitude in its landing approach), and a probable contributing cause was an excessive "sink rate" during the final stages.[34]

All the accidents were attributed to pilot error. According to information from a 727 pilot and the N.T.S.B., pilots were being given only twenty-five hours' training on the aircraft before flying passenger services. Some of the plane's characteristics were shown to need the pilot's special attention. For instance, in most T-tails it could be fatal to "underspool" (that is, run the engines at low revolutions) during a landing approach with landing empennage extended, because of the high sink rate and the fact that it takes between eight and ten seconds for the plane to pick up speed after adjustment of the throttles.

One can understand some early handling problems with entirely new aircraft designs. But why should it require five costly accidents (not to mention the other early T-tail crashes outside the United States) before a problem is mastered? Since the boards of inquiry have so regularly opted for the ever-convenient verdict of "pilot error" in these cases, it is not easy to reopen the argument without fresh evidence. But when one gets a quick succession of crashes in the same type of aircraft, over a relatively short space of time, flown by supposedly experienced profes-

[34] United Air Lines, August 16, 1965, 30 dead; American Airlines, November 8, 1965, 58 dead, 4 injured; United Air Lines, November 11, 1965, 43 dead, 35 injured, 13 saved; All-Nippon Airways, February 4, 1966, 133 dead.

sionals, one suspects that human frailty is not the only weak point in the causal chain.

Generally speaking, these inquiry boards seem anxious only to make some neat attribution of blame. They are thorough up to a point. But there is little attempt by these bodies (so strongly placed because of their impartiality) to help make aviation safer. They will certainly say that a defective bolt in an aileron was the prime cause of a crash. But there is no rigorous pursuit of the reasons why it was there. Neither the designer nor the government inspector nor the manufacturer himself will be given an uncomfortable time in the witness box. If the reader still doubts that this is necessary, let him look again at the accident defects listed on previous pages and note how often the same fault comes up several times from the same manufacturer.

Operational experience has shown that the T-tails are no more unsafe than any other aircraft, *once their handling characteristics are known*. But, when the first one crashed, might it not have been reasonable to expect some useful recommendation from the board of inquiry—that the training period on these new types, let us say, should be doubled before any more passengers were put at risk? This did not happen; nor did it with the second or the third such accident.

The T-tails are clearly of "advanced" design. Immense effort and skill in design went into making them fast and economical. Yet, when we see them in terms of safety, they are primitive. No particular lessons have been learned that one can notice. No imagination about the protection and survival of passengers has been exercised. The persistent drive for economy in "marginal" things has assumed control.

I dealt at some length with the inadequacies of the early models of the Boeing 727 in Chapter 1, in discussing the Salt Lake City crash, with reference principally to the views and experience of those who survived. This, I suggest, is the sort of testimony that should be listened to closely by designers. The designer's decision about what looks right on the drawing board can affect whether such people survive or die. I will recapitulate briefly, since the gaps in safety precautions are similar for all these T-tailed jets. The escape exits are poor. The ventral staircase (the one at the rear, through which most economy-class passengers enter) is of dubious value as an emergency exit. The reasons are obvious. Because of the weight and flight characteristics of the rear engines, the tendency of all these T-tails is to suffer crash damage at the back of the aircraft, thus jamming the exit, or, as happened at Salt Lake, to cause the undercarriage to collapse, thus making the opening of the ventral stairs impossible. The number of exits is not enough to cope with the number of passengers carried, even though these exits do comply with the regulations; with the rapid spread of a fire, and with the combustible and toxic materials in the cabins of most of today's planes, the passenger doesn't stand much chance of survival in a more serious accident. When the collapse of the undercarriage in this aircraft showed that it could instantly cause the fuel lines to be ruptured, the Boeing company was forced to take some steps to strengthen and reroute those lines under the cabin floor so that they were partly shielded by the main *longerons* (longitudinal members of the steel chassis).[35]

The Trident One had its fuel lines outside the main fuselage altogether, but in a shielding; then, in the Trident Two, they were put back inside the fuselage. The Trident Three will shortly appear. Where will they be in this aircraft? Inches inside the fuselage? Or further away from potential damage during a bad landing and right away from undercarriage legs should they penetrate the fuselage? If the designers can solve the problems of speed and fuel economy so

[35] C.A.B. Accident Investigation Report File No. 1–0032, June 7, 1966.

skillfully, the answers to some of the more intractable survival problems should also be within their grasp. But it is important to establish that apostles of greater safety are not demanding the impossible, or even something that is still over the horizon. Practical answers to some of the more elementary kinds of passenger risk have been ready and waiting for years, needing only a manufacturer, and some of his airline customers, with enough interest to apply them. On more advanced problems, a large body of research has been built up, but reports on it too, for the most part, lie on the shelf, collecting dust.

One basic assumption that these reports challenge is that a high rate of death and injury in most air crashes is somehow "inevitable." For example, one can turn to a recent report called "Crash Survival Design Guide," which contains the results of research by a team of experts brought together by the Aviation Safety Engineering and Research Foundation in the United States. One of its conclusions is that "personnel survival is possible even in accidents far more severe than generally assumed by those not acquainted with this field."

One area in which advances in crash safety is urgently needed, it suggests, is in "airframe crashworthiness." The fuselage must form a "protective shell" around the passengers. Proper design can help in this; an excessive increase in weight is not necessary. For instance, one fault common to all commercial aircraft is that the nose structure is relatively soft. If the aircraft strikes the ground at even a slight angle, the underside of the nose tends to buckle, thus forming a scoop, which digs stubbornly into the ground. This greatly increases the deceleration force transmitted to the rest of the airframe, the floor, the seats and the passengers. If the nose was strengthened so that it would slide, this impact force would be reduced.

The death and injury rates also increase sharply with the

breakup of the cabin floor. Research shows that this can be minimized by installing collapsible floor mounts, so that the floor will give way gradually, if at all, or by putting thick padding material between the floor and the cargo hold. The seats themselves must be secured to withstand a crash impact of more than 9 g (that is, nine times the force of gravity). This is the standard laid down by the civil-aviation authorities. Under any impact stronger than this, the seats will break away. But American research has shown that in a large number of accidents the crash force has been between 12 and 15 g. The seats have been uprooted, whereas, if they had been properly secured, lives could have been saved and injuries prevented. Similarly, other research has shown that the human frame can withstand far greater impact force than designers cater for. Tests have shown that a two-inch padding on seat backs and all other potential striking surfaces in the cabin would protect passengers from *most* impact injury in *all* accidents which are conceivably survivable. Where is the aircraft seat today which has any such padding? The whole design conception of aircraft needs to be changed with safety in the forefront. Ductile metals which allow twisting and buckling without rupture could be used more frequently in the fuselage. "The structural shell should be able to deform without fracture as far as possible," says the Flight Safety Foundation.[36] How many modern aircraft could claim to represent any response to this design principle?

The contrasts between expert research and current manufacturing practice are even more marked when we look at emergency exits. The Flight Safety Foundation urges that they should be "equally divided on each side of the fuselage and should not be directly opposite each other." This is obviously good logic, since it would separate people pressing

[36] "Crash Survival Design Guide," Department of Commerce Document AD656 621.

Designed for Disaster

to get out into manageable groups instead of allowing, as now, a crush in one or two places. Other recommendations are that the exits be placed away from fuel-spillage areas; that they should be capable of being opened by rescuers outside; and that the release mechanism can be completely operated within five seconds; and that they should be automatically exploded open before the crashed plane stops and *before* it strikes something that will buckle the fuselage and jam the exits shut. Where is the aircraft now in production which has these features?

The Boeing 747 and similar aircraft soon to appear seem to bring us a new era in flying. Outwardly these vehicles are astonishing and impressive. They are even larger than one could have imagined. Yet, as with every other jet flying today, the safety provision in them seems to be still lagging behind research knowledge.

A new F.A.A. regulation requires a cabin baggage locker for each passenger. This is a wise step. But there has been no updating of seating strengths. First-class seats will be about four in a line; economy-class seats will be nine or ten together. The ten emergency exits are double size; but they still have to be manually opened, and no noticeable safeguards have been built in to prevent them from jamming after an impact. There is no indication that the floors have been strengthened. My view, shared by a number of informed people in aviation, is that aircraft of this huge passenger capacity should not be put into operation until answers to the safety problem are built into them. The methods of evacuating a 747 are by "escape slides," or "chutes." These are in use in today's aircraft, and their history in accidents is a long way from perfection. They have been known *not* to inflate, or to inflate only to be consumed by the fire outside; they have been known to have major faults and not to operate at all; so, with presumably the same possibilities in the 747, the only thing left for the passenger to do is to jump the sixteen feet to the ground. Already there have been broken legs and ankles sustained by people jumping ten feet from a crashed aircraft.

Finally, we have to ask, who can do something about it? The airlines and the manufacturers have the means. But they are evidently too preoccupied with economies of various kinds to be realistically capable, on their own volition, of the vigorous effort required. It comes back in the end to persuasion by the public and, if necessary, coercion by the civil-aviation authorities who fix the regulations. It is they who prescribe these unsatisfactory, minimal seating strengths and the inadequate emergency exits, and it is they who show no sign of having studied the latest research on a dozen other features of passenger protection. These must be updated without delay, before there is any more needless loss of life.

8 · What to Do in an Emergency

It is very important to work out in advance what you will do when the plane hits the ground. Not only will this enable you to rationally choose the best alternative, but the effort of working out a plan will keep your mind active during the crisis and take it off the terror of the events at hand.

1. As soon as an emergency has been declared, remove all sharp, pointed and metallic objects from your pockets and put them in the pocket of the seat in front of you. Such objects can cause major bodily injuries on impact.

2. Remove the life jacket from the container under your seat and put it on. It is amazing how many times pilots make emergency landings on what they assume to be ground, only to find themselves on a lake or river. With your life jacket on, you are ready for such a situation. Do not under any circumstances inflate the life jacket before emerging from the plane.

3. Straighten the back of your seat and sit in the middle of it. If possible, put blankets or pillows between the sides of your body and the arm rests. This way you will be more likely to avoid injuring yourself against the seats.

4. Tighten your seat belt. When the plane hits, you will probably be thrown downward, thus slackening the belt some and inviting your sliding out from under it. A tight belt will correct this problem.

5. Put your feet in front of you on the floor, making sure they are not under your seat or the seat in front. This

way you will have the best chance of avoiding leg injuries, although in the event of a general seat collapse, it will be very difficult to avoid injury. However, if your feet are under the seat in front, you may help it collapse if there are upward forces which move your feet up against it. The seats have less ability to resist upward forces than any other kind.

6. The most important thing you can do for your personal safety is to place some padding, cushions or blankets, around your head when the final "brace for impact" instructions are given.

7. While waiting for the "brace for impact" order read and reread the instructions noting your nearest way of escape, work out how to operate the exit and read the instructions on how to inflate the life rafts. When you are fully clear in your mind as to just exactly how you are going to effect your escape, work out an alternative route (your original route could be blocked by jamming, fire, explosion or objects outside the aircraft). Make alternative plans for an exit on the opposite side of the aircraft from the original. Finally, if there is still time, place handbags, baggage, briefcases, et cetera, into the compartment under the seat which held your life jacket.

8. When the order "brace for impact" is given, you'll still have about thirty seconds, provided that the crew has been properly trained. Cover your head with the blanket and/or pillow, bend forward placing forehead on knees and hold the padding (blanket and cushion) with your arms, fingers interlocked, over the back of your neck, elbows pulled in close to your ears, and with your arms push your face onto your knees. Try and relax most of your body by placing pressure on the back of your head. At the same time place pressure on your legs to hold them in the straight up-and-down position.

9. There will most likely be *more than one impact.*

Do not move from this "brace for impact" position until the plane is very evidently close to or practically stopping. If the plane is ditching remember that it could "nose in" after the initial impact. If on land, something large and solid could be in its crash path.

10. When the plane has stopped *and not before* (unless fire has already taken hold and is in your immediate vicinity), sit up, undo your seat belt, discard the padding materials and quickly make your planned escape. Remember your speed in getting away from the crashed aircraft could mean the difference between life and death. Leave your belongings behind (the airline's insurance covers them).

11. Once out of the aircraft move well away, especially if you are injured or if there is any sign of smoke or fire.

DATE DUE

JOHN GODSON was born in Australia and has had a distinguished career in broadcast journalism in that country, England, and Ireland. For several years he was a news director and producer for the BBC television network in London. He now lives in Switzerland with his wife.